The Public Health Emergency Medical Countermeasures Enterprise

Innovative Strategies to Enhance Products from Discovery Through Approval

Workshop Summary

Theresa Wizemann, Clare Stroud, and Bruce M. Altevogt,
Rapporteurs

Forum on Medical and Public Health Preparedness for Catastrophic Events

Forum on Drug Discovery, Development, and Translation

Board on Health Sciences Policy

INSTITUTE OF MEDICINE
OF THE NATIONAL ACADEMIES

THE NATIONAL ACADEMIES PRESS
Washington, D.C.
www.nap.edu

THE NATIONAL ACADEMIES PRESS • 500 Fifth Street, N.W. • Washington, DC 20001

NOTICE: The project that is the subject of this report was approved by the Governing Board of the National Research Council, whose members are drawn from the councils of the National Academy of Sciences, the National Academy of Engineering, and the Institute of Medicine. The members of the committee responsible for the report were chosen for their special competences and with regard for appropriate balance.

This project was supported by contracts between the National Academy of Sciences and the American College of Emergency Physicians, the American Hospital Association, the American Medical Association, the American Nurses Association, the Association of State and Territorial Health Officials, the Centers for Disease Control and Prevention (Contract No. 200-2005-13434 TO #6), the Department of the Army (Contract No. W81XWH-08-P-0934), the Department of Health and Human Services' Agency for Healthcare Research and Quality (Contract No. HHSP233200800498P), the Department of Health and Human Services' National Institutes of Health (Contract No. N01-OD-4-2139 TO #198), the Department of Health and Human Services' Office of the Assistant Secretary for Preparedness and Response (Contract Nos. HHSP233200900680P, HH5P23320042509X1), the Department of Homeland Security's Office of Health Affairs (Contract No. HSHQDC-07-C-00097), the Department of Homeland Security's Federal Emergency Management Agency (Contract No. HSFEHQ-08-P-1800), the Department of Veterans Affairs (Contract No. V101(93)P-2136 TO #10), the Emergency Nurses Association, the National Association of Chain Drug Stores, the National Association of County and City Health Officials, the National Association of Emergency Medical Technicians, the Pharmaceutical Research and Manufacturers of America, The Robert Wood Johnson Foundation, and the United Health Foundation. The views presented in this publication are those of the editors and attributing authors and do not necessarily reflect the views of the organizations or agencies that provided support for this project.

International Standard Book Number-13: 978-0-309-15024-8
International Standard Book Number-10: 0-309-15024-8

Additional copies of this report are available from The National Academies Press, 500 Fifth Street, N.W., Lockbox 285, Washington, DC 20055; (800) 624-6242 or (202) 334-3313 (in the Washington metropolitan area); Internet, http://www.nap.edu.

For more information about the Institute of Medicine, visit the IOM home page at: **www.iom.edu.**

Printed in the United States of America

Suggested citation: IOM (Institute of Medicine). 2010. *The Public Health Emergency Medical Countermeasures Enterprise: Innovative Strategies to Enhance Products from Discovery Through Approval: Workshop Summary.* Washington, DC: The National Academies Press.

"Knowing is not enough; we must apply.
Willing is not enough; we must do."
—Goethe

INSTITUTE OF MEDICINE
OF THE NATIONAL ACADEMIES

Advising the Nation. Improving Health.

THE NATIONAL ACADEMIES

Advisers to the Nation on Science, Engineering, and Medicine

The **National Academy of Sciences** is a private, nonprofit, self-perpetuating society of distinguished scholars engaged in scientific and engineering research, dedicated to the furtherance of science and technology and to their use for the general welfare. Upon the authority of the charter granted to it by the Congress in 1863, the Academy has a mandate that requires it to advise the federal government on scientific and technical matters. Dr. Ralph J. Cicerone is president of the National Academy of Sciences.

The **National Academy of Engineering** was established in 1964, under the charter of the National Academy of Sciences, as a parallel organization of outstanding engineers. It is autonomous in its administration and in the selection of its members, sharing with the National Academy of Sciences the responsibility for advising the federal government. The National Academy of Engineering also sponsors engineering programs aimed at meeting national needs, encourages education and research, and recognizes the superior achievements of engineers. Dr. Charles M. Vest is president of the National Academy of Engineering.

The **Institute of Medicine** was established in 1970 by the National Academy of Sciences to secure the services of eminent members of appropriate professions in the examination of policy matters pertaining to the health of the public. The Institute acts under the responsibility given to the National Academy of Sciences by its congressional charter to be an adviser to the federal government and, upon its own initiative, to identify issues of medical care, research, and education. Dr. Harvey V. Fineberg is president of the Institute of Medicine.

The **National Research Council** was organized by the National Academy of Sciences in 1916 to associate the broad community of science and technology with the Academy's purposes of furthering knowledge and advising the federal government. Functioning in accordance with general policies determined by the Academy, the Council has become the principal operating agency of both the National Academy of Sciences and the National Academy of Engineering in providing services to the government, the public, and the scientific and engineering communities. The Council is administered jointly by both Academies and the Institute of Medicine. Dr. Ralph J. Cicerone and Dr. Charles M. Vest are chair and vice chair, respectively, of the National Research Council.

www.national-academies.org

Workshop Planning Committee*

GAIL CASSELL (*Chair*), Eli Lilly and Company, Indianapolis, IN
PHYLLIS ARTHUR, Biotechnology Industry Organization, Washington, DC
CAPTAIN KENNETH COLE, Office of Assistant Secretary of Defense for Health Affairs, Department of Defense, Washington, DC
ALEXANDER GARZA, Department of Homeland Security, Washington, DC
JOHN GRABENSTEIN, Merck & Co., Inc., Whitehouse Station, NJ
ELIN GURSKY, Analytic Services Inc., Arlington, VA
TIL JOLLY, Department of Homeland Security, Washington, DC
ROBERT KADLEC, PRTM Management Consultants, Washington, DC
GEORGE KORCH, Department of Health and Human Services, Washington, DC
MICHAEL KURILLA, National Institute of Allergy and Infectious Diseases, Bethesda, MD
NICOLE LURIE, Department of Health and Human Services, Washington, DC
BORIS LUSHNIAK, Food and Drug Administration, Rockville, MD
MONIQUE MANSOURA, Department of Health and Human Services, Washington, DC
PETER PALESE, Mount Sinai School of Medicine, New York
RON SALDARINI, Biological Initiatives, Mahwah, NJ
DANIEL SOSIN, Centers for Disease Control and Prevention, Atlanta, GA
JANET TOBIAS, Ikana Media, New York
ERIC TONER, University of Pittsburgh, Pittsburgh, PA
STEPHANIE ZAZA, Centers for Disease Control and Prevention, Atlanta, GA

---

* IOM planning committees are solely responsible for organizing the workshop, identifying topics, and choosing speakers. The responsibility for the published workshop summary rests with the workshop rapporteurs and the institution.

IOM Staff

BRUCE ALTEVOGT, Preparedness Forum Director
ROBERT GIFFIN, Drug Forum Director (until March 2010)
CLARE STROUD, Program Officer
ANDREW POPE, Director, Board on Health Sciences Policy
AMY PACKMAN, Administrative Assistant
ALEX REPACE, Senior Program Assistant

Forum on Medical and Public Health Preparedness for Catastrophic Events[*]

LEWIS GOLDFRANK (*Chair*), New York University Medical Center, New York
DAMON ARNOLD, Association of State and Territorial Health Officials, Arlington, VA
GEORGES BENJAMIN, American Public Health Association, Washington, DC
D. W. CHEN, Office of Assistant Secretary of Defense for Health Affairs, Department of Defense, Washington, DC
ROBERT DARLING, Uniformed Services University, Bethesda, MD
VICTORIA DAVEY, Department of Veterans Affairs, Washington, DC
JEFFREY DUCHIN, Seattle & King County and University of Washington, Seattle
ALEXANDER GARZA, Department of Homeland Security, Washington, DC
LYNN GOLDMAN, Johns Hopkins Bloomberg School of Public Health, Baltimore, MD
DAVID HENRY, National Governors Association, Washington, DC
JACK HERRMANN, National Association of County and City Health Officials, Washington, DC
KEITH HOLTERMANN, Federal Emergency Management Agency, Washington, DC
JAMES JAMES, American Medical Association, Chicago, IL
JERRY JOHNSTON, National Association of Emergency Medical Technicians, Mt. Pleasant, IA
ROBERT KADLEC, PRTM Management Consultants, Washington, DC
BRIAN KAMOIE, The White House, Washington, DC
LYNNE KIDDER, Business Executives for National Security, Washington, DC
MICHAEL KURILLA, National Institute of Allergy and Infectious Diseases, Bethesda, MD

---

[*] IOM forums and roundtables do not issue, review, or approve individual documents. The responsibility for the published workshop summary rests with the workshop rapporteurs and the institution.

JAYNE LUX, National Business Group on Health, Washington, DC
ANTHONY MACINTYRE, American College of Emergency Physicians, Washington, DC
ANGELA MCGOWAN, Robert Wood Johnson Foundation, Princeton, NJ
MARGARET MCMAHON, Emergency Nurses Association, Williamstown, NJ
ERIN MULLEN, Pharmaceutical Research and Manufacturers of America, Washington, DC
GERALD PARKER, Office of the Assistant Secretary for Preparedness and Response, Department of Health and Human Services, Washington, DC
CHERYL PETERSON, American Nurses Association, Silver Spring, MD
SALLY PHILLIPS, Agency for Healthcare Research and Quality, Rockville, MD
STEVEN PHILLIPS, National Library of Medicine, Bethesda, MD
EDITH ROSATO, National Association of Chain Drug Stores Foundation, Alexandria, VA (since July 2009)
PHILLIP SCHNEIDER, National Association of Chain Drug Stores Foundation, Alexandria, VA (until July 2009)
ROSLYNE SCHULMAN, American Hospital Association, Washington, DC
DANIEL SOSIN, Centers for Disease Control and Prevention, Atlanta, GA
SHARON STANLEY, American Red Cross, Washington, DC
ERIC TONER, University of Pittsburgh Medical Center, Pittsburgh, PA
REED TUCKSON, UnitedHealth Group, Minneapolis, MN
MARGARET VANAMRINGE, The Joint Commission, Washington, DC

IOM Staff

BRUCE ALTEVOGT, Project Director
CLARE STROUD, Program Officer
ANDREW POPE, Director, Board on Health Sciences Policy
ALEX REPACE, Senior Program Assistant

Forum on Drug Discovery, Development, and Translation*

* IOM forums and roundtables do not issue, review, or approve individual documents. The responsibility for the published workshop summary rests with the workshop rapporteurs and the institution.

FREDA LEWIS-HALL, Pfizer Inc., New York
WILLIAM MATTHEW, National Institute of Neurological Disorders and Stroke, Bethesda, MD
MARK McCLELLAN, Brookings Institution, Washington, DC
CAROL MIMURA, University of California–Berkeley, CA
JOHN ORLOFF, Novartis Pharmaceuticals Corporation, East Hanover, NJ
AMY PATTERSON, National Institutes of Health, Bethesda, MD
JANET SHOEMAKER, American Society for Microbiology, Washington, DC
LANA SKIRBOLL, National Institutes of Health, Bethesda, MD
NANCY SUNG, Burroughs Wellcome Fund, Research Triangle Park, NC
JORGE TAVEL, National Institute of Allergy and Infectious Diseases, Bethesda, MD
JANET TOBIAS, Ikana Media, New York
JOANNE WALDSTREICHER, Johnson & Johnson Pharmaceutical Research and Development, LLC, Raritan, NJ
JANET WOODCOCK, U.S. Food and Drug Administration, Rockville, MD
RAYMOND WOOSLEY, The Critical Path Institute, Tucson, AZ

IOM Staff

ANNE CLAIBORNE, Director (since April 2010)
ROBERT B. GIFFIN, Director (until March 2010)
REBECCA A. ENGLISH, Research Associate
YEONWOO LEBOVITZ, Program Associate
GENEA S. VINCENT, Senior Program Assistant

Reviewers

This report has been reviewed in draft form by individuals chosen for their diverse perspectives and technical expertise, in accordance with procedures approved by the National Research Council's Report Review Committee. The purpose of this independent review is to provide candid and critical comments that will assist the institution in making its published report as sound as possible and to ensure that the report meets institutional standards for objectivity, evidence, and responsiveness to the study charge. The review comments and draft manuscript remain confidential to protect the integrity of the process. We wish to thank the following individuals for their review of this report:

Susan R. Cooper, Tennessee Department of Health
Joseph A. DiMasi, Tufts Center for the Study of Drug Development
Philip K. Russell, Department of Defense (retired)
Patrick J. Scannon, XOMA, Ltd.
P. Roy Vagelos, Merck & Co., Inc. (retired)

Although the reviewers listed above have provided many constructive comments and suggestions, they did not see the final draft of the report before its release. The review of this report was overseen by **Leslie Z. Benet.** Appointed by the Institute of Medicine, he was responsible for making certain that an independent examination of this report was carried out in accordance with institutional procedures and that all review comments were carefully considered. Responsibility for the final content of this report rests entirely with the authoring committee and the institution.

Contents

INTRODUCTION[1]

"We are launching a new initiative that will give us the capacity to respond faster and more effectively to bioterrorism or an infectious disease—a plan that will counter threats at home and strengthen public health abroad."

—President Barack Obama, 2010 State of the Union Address

Safe and effective medical countermeasures, including vaccines, drugs, and diagnostics, are critical for responding to large-scale public health emergencies. Such situations, be they natural (e.g., pandemic influenza) or man-made (e.g., terrorism), have the potential to rapidly overwhelm public health and medical systems. America's national security depends on having appropriately licensed chemical, biological, radiological, and nuclear medical countermeasures in its arsenal of defenses.

The Public Health Emergency Medical Countermeasures Enterprise (PHEMCE or countermeasures enterprise)[2] encompasses diverse

---

[1] The workshop was organized by an independent planning committee whose role was limited to the identification of topics and speakers. This workshop summary was prepared by the rapporteurs as a factual summary of the presentations and discussions that took place at the workshop. Statements, recommendations, and opinions expressed are those of individual presenters and participants, and are not necessarily endorsed or verified by the Forums or the National Academies, and should not be construed as reflecting any group consensus. Furthermore, although the current affiliations of speakers and panelists are noted in the report, many qualified their comments as being based on personal experience over the course of a career, and not being presented formally on behalf of their organization (unless specifically noted).

[2] The PHEMCE, led by the Health and Human Services (HHS) Office of the Assistant Secretary for Preparedness and Response, includes the Centers for Disease Control and Prevention, the Food and Drug Administration, and the National Institutes of Health. Interagency partners include the Department of Homeland Security, the Department of Defense, the Department of Veterans Affairs, and the Department of Agriculture. The

partners from across federal, state, and local governments, industry, and academia. Despite its successes, certain structural, strategic, and technical elements of the countermeasures enterprise continue to impede research, development, and production of medical countermeasures. The National Institutes of Health (NIH) and the Department of Defense (DoD) support much of the basic research in the relevant health and disease areas. However, this research is not always aligned with the top priorities identified based on threat assessments, which limits the number of discoveries that are applicable for further development as medical countermeasures. Once potential candidates for advanced development are identified, they are often not yet at a stage of development where they can be handed off to the Biomedical Advance Research and Development Authority (BARDA)[3] in the U.S. Department of Health and Human Services (HHS). Furthermore, because the commercial market is limited for most medical countermeasures, it can be difficult to engage private-sector pharmaceutical and biotechnology companies to participate in the development and manufacturing of these products.

To begin to address the efficiency and effectiveness issues of the PHEMCE, on December 1, 2009, HHS Secretary Kathleen Sebelius charged the "Office of the Assistant Secretary for Preparedness and Response [ASPR] to lead a review of its entire public health countermeasures enterprise, to be completed in the first quarter of next year." Subsequently, in response to a request from the Assistant Secretary, the Institute of Medicine's (IOM's) Forum on Medical and Public Health Preparedness for Catastrophic Events and Forum on Drug Discovery, Development, and Translation jointly convened a workshop on February 22–24, 2010, titled The Public Health Emergency Medical Countermeasures Enterprise: Innovative Strategies to Enhance Products from Discovery Through Approval. The workshop was designed to examine federal policies and activities that affect medical countermeasure discovery, development, and approval, and to explore potential opportunities to enhance the countermeasures enterprise by

---

PHEMCE mission is to optimize national preparedness for public health emergencies, specifically by the creation, stockpiling, and use of medical countermeasures.

[3] BARDA's mission is to provide countermeasures for chemical, biological, radiological, and nuclear threats, pandemic influenza, and emerging infectious diseases through product requirement setting, product development, stockpile acquisition/building, manufacturing infrastructure building, and product innovation. BARDA resides within ASPR, manages the PHEMCE, and has the procurement authority for Project BioShield acquisitions using the Special Reserve Fund (http://www.hhs.gov/aspr/barda/index.html).

evaluating existing models or systems having similar goals of developing medical products with low commercial viability (Box 1).[4]

BOX 1
Workshop Objectives

- Identify and discuss strategies to optimize the federal public health emergency medical countermeasures enterprise, and explore resources and/or other supporting components needed for accomplishing goals of countermeasure discovery, development, approval, and production.
- Examine strategies to further enhance the translation of early phase investments in basic science into potential public health interventions.
- Identify and discuss models for enhancing current partnerships and establishing new ones among federal programs, innovators, and the commercial marketplace to enhance our nation's capabilities to meet public health emergency preparedness goals.
- Consider market forces acting on the advanced development biodefense community (pharma/biotech) that incentivize/ disincentivize efforts to develop and license products in support of the national response.
- Examine ways the regulatory oversight process for public health emergency medical countermeasures might evolve and identify ways to enable more efficient approval and use.
- Review the innovative approaches being used to advance drug development for orphan diseases (i.e., rare, neglected, or tropical diseases) or any other area that does not have a ready and sustainable commercial market (e.g., oncology therapeutics) and identify the shared challenges and opportunities for strategies that might be adopted by the countermeasures enterprise.

About This Summary

This document highlights and summarizes the work presented at the workshop with the hope that this information will help federal officials to conduct a thorough review of the pipeline through approval spectrum of our national programs and to assist in the ultimate goal of improving the efficiency and effectiveness of the countermeasures enterprise. Whenever possible, unique ideas or concepts presented at the meetings are attributed in this report to the individual who first advanced those concepts. In situations where many attendees made similar points, the

[4] Audio files, slides, and the meeting transcript are available for download via the Preparedness Forum's website, http://www.iom.edu/preparednessforum.

recurring themes are identified. The final section of the summary lists a number of suggestions for improving the medical countermeasures enterprise, including a number of suggestions focused on countermeasure regulation and licensure. They are compiled here as part of the factual summary of the workshop, and should not be construed as reflecting consensus or endorsement by the workshop, the Forums, or the National Academies. Investigating details about the feasibility and implementation of these ideas were beyond the scope of the workshop and this summary.

Charge to Workshop Participants

In her opening comments and charge to the workshop participants, the HHS Assistant Secretary for Preparedness and Response, Nicole Lurie, said that time and time again, it has been apparent that the United States does not necessarily have the countermeasures needed to respond to a public health emergency, regardless of whether it is natural or initiated by humans. Although signs of progress have been apparent in recent years, much more work is needed to protect the nation against the range of potential threats. Using the recent H1N1 influenza pandemic as an example, she also noted that even when countermeasures are available, low levels of public acceptance of the countermeasure can inhibit an effective response, and significant public education efforts may be required.

A primary goal of the end-to-end review of the public health countermeasures enterprise is to understand, in enough detail to be actionable, the challenges related to the current approach to develop countermeasures and the opportunities to improve them. Many of the challenges are already well known. Lurie urged workshop participants to be frank and forthcoming in offering creative solutions, calling for a very granular and specific focus on understanding the needs and developing strategies for systemic change. ASPR is seeking to understand how the incentive structures, policies, and procedures are, or are not, aligned with the needs of the pharmaceutical and biotechnology industries, the United States government, and the American people. The discussions at the workshop also helped inform the deliberations that were under way by the National Biodefense Science Board (NBSB),[5] which was charged by

---

[5] The National Biodefense Science Board was created under the authority of the Pandemic and All-Hazards Preparedness Act (Public Law 109-417) "to provide expert advice to the Secretary on scientific, technical and other matters of special interest to HHS regarding current and future chemical, biological, nuclear and radiological agents, whether naturally occurring, accidental or deliberate. The Board may also provide advice and

HHS to conduct a parallel examination of the related strategic management, leadership and accountability structure of the PHEMCE (NBSB, 2010a).

BACKGROUND

To aid their review, ASPR commissioned a set of briefings from PRTM Management Consultants that was presented at the workshop (Box 2). Two of these briefings, #1 and #3, were developed into white papers that serve as Appendixes D and E of this workshop summary.

BOX 2
Highlights of Commissioned White Papers

Case Studies of the HHS Medical Countermeasure Programs: Briefing #1

Select case studies of Department of Health and Human Services (HHS) medical countermeasure programs were examined (anthrax, smallpox, hematopoietic acute radiation syndrome, viral hemorrhagic fevers, broad-spectrum antibiotics for bacterial threats) to evaluate: What were the successful elements of each program? What were the setbacks, real or perceived failures, of each program? What improvements could be made to improve future programs?

Although there is not one event/characteristic that portends failure or guarantees success, there are shared risks identified in each case study, and some common factors that appear to increase the likelihood of success.

Three factors that impact successful drug development are a failure in efficacy, a failure in safety (accounting for about two-thirds of failures), and failure in commercial considerations (e.g., cost to bring the product to market, perceived profitability of the product).

Common factors of successful programs are strong leadership from the top, realistic expectations, experienced people, mature organizations, and adequate resources.

guidance to the Secretary on other matters related to public health emergency preparedness and response" (http://www.hhs.gov/aspr/omsph/nbsb/).

Optimizing the Medical Countermeasure Product Pipeline from the Science Base Through Advanced Development: Briefing #2

This briefing addresses how the product pipeline can be increased to improve the chances of producing approved products for the Public Health Emergency Medical Countermeasures Enterprise (PHEMCE), identifying challenges and looking to other programs for solutions.

Although no single model or specific solution ensures success, observations from comparative research and development models suggest that better management structures, strategic decision making, better definition of requirements, target product profiles, and defined metrics of success may increase the PHEMCE pipeline of candidate products. Successful models have incorporated partnerships to optimize limited funding, market assurance, and pursuing products with multiuse potential.

Synthesis of Business Models and Economic and Market Incentives for Vaccines and Therapeutics: Briefing #3

Increasing the level and mix of pharmaceutical and biotechnology company engagement can bring critical knowledge and experience to the PHEMCE. Based on interviews and the literature, three major deterrents to industry engagement in medical countermeasures development were identified: requirements (insufficient granularity/clarity about what the government wants, what companies are being asked to make, how it will be sold); return on investment (unpredictable, unsustainable market); and uncertainty in the regulatory pathway.

This briefing explored multiple push and pull incentives for attracting industry participation that have been proposed or implemented in other contexts, but have not yet been applied to medical countermeasures development. No one push or pull incentive is sufficient to attract experienced companies to participate in medical countermeasures development. Similarly, there is no "silver bullet" combination of incentives. The right response depends on context.

A report by the NBSB titled *Optimizing Industrial Involvement with Medical Countermeasure Development* was also presented as background for the discussions. John Grabenstein of Merck Vaccines, who is a member of the NBSB, said numerous chemical, biological, radiological, and nuclear countermeasures are still needed beyond those licensed medical countermeasures currently available in the Strategic National Stockpile (NBSB, 2010b). The Project BioShield Act provided for a procurement fund to foster the development of medical products that did not yet exist. Although subsequent legislation attempted to target resources for the advanced development of countermeasures, this funding has never been adequate. Although it is important to ensure the procurement resources remain available, Grabenstein explained, far greater resources are currently required to ensure necessary countermeasures research and development.

In describing the findings from the NBSB report, Grabenstein said the U.S. government's medical countermeasures enterprise has made several important advances in improving the environment for countermeasure development, including the creation of BARDA, the option for an Emergency Use Authorization (EUA),[6] and the Animal Rule,[7] as well as the new HHS and DoD commitment toward an "Integrated Portfolio," and the PHEMCE holding stakeholder meetings and workshops. However, barriers hindering industry involvement in the development of countermeasures remain, including inadequate and inconsistent funding, opportunity costs (e.g., distractions from other company priorities), economics (e.g., financial margins and low volumes), uncertain regulatory pathways, finite human capital (a limited number of people having the necessary, specialized skill sets), the complexity of working with multiple federal agencies, inadequate federal government understanding of the commercial biopharmaceutical enterprise, and the use of an acquisition system that was originally created to procure complex mechanical equipment such as aircraft, vehicles, and ships (NBSB, 2010b). To begin to address these issues, the NBSB report offers a list of eight specific recommendation for the government, which are further detailed in the full report (Box 3).

---

[6] Under Section 564 of the Federal Food, Drug, and Cosmetic Act, as amended by Project BioShield Act of 2004, the Commissioner of the Food and Drug Administration may authorize the use of an unapproved medical product, or an unapproved use of an approved medical product, during a declared emergency involving a heightened risk of attack on the public or U.S. military forces, or a significant potential to affect national security (http://www.fda.gov/RegulatoryInformation/Guidances/ucm125127.htm).

[7] The Animal Rule allows for the approval of drugs (21 C.F.R. 314.600) or biological products (21 C.F.R. 601.90) based on evidence of effectiveness from studies in animals under certain conditions when human efficacy studies are not ethical or feasible.

BOX 3
Specific Recommendation of the National Biodefense Science Board to the U.S. Government

1. To harness the national industrial base, the U.S. Congress and the Executive Branch must provide adequate, consistent funding (for both advanced development and procurement).
2. The U.S. government must accelerate the pace of medical countermeasure development and acquisition and optimize distribution methods.
3. The U.S. government must centralize its leadership for medical countermeasure development, procurement, and approval.
4. The U.S. government must demonstrate long-term commitment to its industry collaborators.
5. The U.S. government must create, sustain, and enhance innovative partnerships with private industry.
6. The U.S. government should expand medical countermeasure markets to include international partners, state, local, and tribal governments, laboratorians, and first responders in each of these sectors.
7. The U.S. government must do a better job of preparing for emergencies that can be anticipated.
8. Various departments, agencies, and entities of the U.S. government must act in concert to ensure success.

SOURCE: NBSB (2010b).

The Public Health Perspective on Medical Countermeasure Development, Acquisition, and Use

A key challenge for the countermeasures enterprise is how to achieve the greatest health impact in the face of diminishing resources. Thomas Frieden, director, Centers for Disease Control and Prevention (CDC), said an effective response starts with several basic principles, as follows:

- **Define what is needed.** Identify and characterize the threats; identify the at-risk groups and the specific needs of different at-risk subgroups (e.g., pediatric use); determine if new countermeasures are needed; and interface with the intelligence community. Defining what is needed involves a combination of pathogenesis, pathophysiology, the likelihood of use, and the likelihood of dispersal.
- **Decide what to make, and make it.** Assess countermeasure availability; secure EUA as needed; develop stockpiling, distribution, and dispensing logistics; plan for countermeasure use and response; and secure licensure. This will require significant and consistent

investments, and a consistent way to work with industry productively, collaboratively, and perhaps most important, predictably.

- **Ensure that the countermeasures that are developed reach the people who need them most, using everyday systems that can be scaled up.** This may require investing in the establishment or enhancement of more everyday systems (e.g., laboratory, epidemiological, vaccination, or healthcare systems). In this regard, Frieden cited the public health response to the recent H1N1 influenza pandemic. Over 100 million doses of vaccine were distributed with next-day delivery to more than 70,000 sites for vaccination. Despite several recalls, the distribution system for H1N1 vaccination worked extremely well because it used the infrastructure of the Vaccines for Children Program. In contrast, there were significant challenges with the distribution of antiviral medications because the public health system does not have an everyday route for dissemination.
- **Monitor the countermeasures and communicate with the public.** Assess effectiveness, determine if supplies are sufficient to meet demand, determine how to increase demand to improve protection of the public, identify and interpret adverse effects, and look for changes in susceptibility of the pathogen to the countermeasure. Public acceptance of countermeasures depends on monitoring safety signals, analyzing risk, and communicating results frequently.

Going forward, Frieden said, better countermeasure delivery will require better intelligence about the presence, modification, and weaponization of different agents; storage and deployment logistics, evidence-based clinical recommendations and algorithms for use; and laboratory capacity that can adapt to the unexpected.

The FDA Perspective on the Countermeasures Enterprise: Moving Forward

In his keynote address to the workshop, Jesse Goodman, chief scientist and deputy commissioner for science and public health (acting) of the Food and Drug Administration (FDA) emphasized that the time is right for action on medical countermeasures and pandemic preparedness. The public health and national security needs are clear; there are multiple insights from the accomplishments and limitations of Project BioShield and from the experiences with 2009 H1N1 influenza; the public, policy makers, and the administration are interested; and there is bipartisan engagement and collaboration across agencies.

In that regard, Goodman highlighted the FDA's February 24, 2010, announcement with the NIH of a new partnership to advance translation of innovations from basic science to products, including a focus on regulatory science. The agencies will establish a joint leadership council, and jointly issue a Request For Applications with the intent of awarding $6.75 million for research on novel technologies and approaches applicable to the development and regulatory review of medical products. Going forward, Goodman said, FDA is focusing on the following four key principles:

- **End-to-end partnering,** including highly interactive and collaborative engagement and outcomes-oriented management. This means defining how products will be used up front, determining the pathways necessary to evaluate and regulate the product, and identifying scientific gaps. Making regulatory requirements clear is needed to reduce uncertainty. Oversight and review of progress at high levels is also necessary.
- **Increased attention to regulatory science,** to expand agency capacity and knowledge and thereby enhance the quality and integrity of FDA decision making. Develop, assess, and provide tools, methods, models, standards, guidance, and pathways to evaluate product safety, efficacy, and quality (e.g., biomarkers; surrogate endpoints; adaptive and other flexible clinical trial designs; rapid scale-up of production; and rapid methods to assess purity, potency, quality, and contamination). Key elements include leadership and coordination within the agency, training and development of FDA staff, and targeted research within the agency.
- More agile **platform and multiuse technologies** (e.g., vaccine, diagnostic, or monoclonal platforms) that can be rapidly adaptable to address new pathogens. (Goodman noted that platforms will not perform for all pathogens and diseases, and concrete experience with real products is needed to provide enhanced predictability of results and reduce regulatory requirements.)
- **Policies that meet public health needs.** For example, although the EUA is a public health success, it can be cumbersome; the Animal Rule needs to reexamined in light of experience and scientific needs and realities; and consideration of accelerated approval approaches needs to be expanded. Are there other approaches or statuses short of full approval that should be considered?

Government Procurement of Science

Michael Kurilla, director of the Office of Biodefense Research Affairs at the National Institute of Allergy and Infectious Diseases (NIAID), explained that from the NIH perspective, there are three basic mechanisms for procuring science, with increasing focus and control. The first mechanism is geared toward increasing basic science knowledge and generating novel concepts, and the primary mechanism is *grants*. When the intent is vetting concepts (i.e., reduction to practice to demonstrate feasibility), the mechanism usually involved is a *cooperative agreement*, including small business grants and technology-transfer arrangements. When the goal is to take a product forward, from target identification to lead to candidate to human testing, the NIH relies on *contracts* with defined deliverables.

In addition to funding, NIH provides a number of services. Specialized services are available as needed and include, for example, sequencing, reagents, screening, animal model development, and containment. Gap-filling services are focused efforts to advance products, and involve traditional preclinical and clinical drug and vaccine development activities.

To facilitate the discussions, Kurilla offered a quick review of programmatic terminology (Table 1).

TABLE 1 National Institutes of Health Programmatic Terminology

Term	Activities	Management and Review
Project	Single effort focused on a specific candidate countermeasure against a specific agent	Success in meeting milestones and time lines
Portfolio	Focused effort typically organized around a single threat agent with multiple countermeasures	Adequacy of individual projects to cover the range of desired candidates, technical approaches, and developmental maturity
Program	Overall effort focused on multiple countermeasures against multiple threats	Progress across total threat space with emphasis on desired approaches

PARTNERS IN A SINGLE MISSION, DIVERSE CONCERNS AND CHALLENGES

Over the course of the workshop, participants highlighted some of the challenges, gaps, and barriers facing those involved in the countermeasures enterprise. While by no means a comprehensive review, these are some of the more pressing concerns that informed the subsequent discussions on optimizing the countermeasures enterprise. A vast array of structural, strategic, technical, financial, and even cultural elements are involved in the research, development, production, and deployment of medical countermeasures for public health emergencies.

The Growing Threat of Bioweapons

D. A. Henderson, former director of the Office of Public Health Emergency Preparedness, and distinguished scholar at the Center for Biosecurity of University of Pittsburgh Medical Center (UPMC), stressed that countermeasures development needs to be approached with a real sense of urgency, noting that the situation with regard to anthrax is not much better than it was 8 years ago. Today, he said, we do not have that sense of real urgency we felt after 9/11, and yet there is an equal likelihood that an event could occur tomorrow.

Compared to nuclear and other weapons technology, bioterrorism is "relatively easy." In a recent editorial, former senators Bob Graham and Jim Talent said they believed it is unlikely that the United States can ever prevent bioterrorism (Graham and Talent, 2009). Rather, the senators stressed that America's best long-term strategy for biodefense is redefining its prevention efforts, striving to reach a level of preparedness that effectively removes bioweapons from the category of weapons of mass destruction (WMDs).

With this in mind, Graham, Talent, and Randy Larson, who served as executive director of the Commission on the Prevention of Weapons of Mass Destruction Proliferation and Terrorism, formed the Bipartisan WMD Terrorism Research Center, a 501(c)(3) organization operational as of March 1, 2010. The organization's primary focus will be education, ensuring that those in leadership positions in the federal government understand the imminent threat that biological weapons present.

Larson urged workshop participants working with and within government on the countermeasures enterprise to request the Department of Homeland Security Office of Science and Technology's population threat assessment briefing. The threat is real, Larson said, but people do not have a full understanding of that threat and therefore do not always apply themselves fully toward solutions.

Gaps and Barriers to International Collaboration

Maria Julia Marinissen of ASPR reminded participants that the threat of terrorism with chemical, biological, radiological, and nuclear agents, and the spread of pandemics and other potential emerging infectious diseases, are global issues. The United States has experienced a steadily increasing demand for the supply of medical countermeasures to foreign countries. However, it is virtually impossible for a single country to fund research and development, acquisition, and stockpiling programs for medical countermeasures for all, or even most, threat agents. A global infrastructure for countermeasures is needed.

In its recommendations to the new administration, the IOM Committee on the U.S. Commitment to Global Health stated that "good health is a necessary condition for economic development and global prosperity" and concluded that this country can improve the lives of millions around the world, while reflecting America's values and protecting and promoting the nation's interests (IOM, 2008). However, the United States cannot become the world's provider and pharmacy for medical countermeasures. A sustainable U.S. infrastructure depends on a larger marketplace for these products.

Over the past 2 years, Marinissen said, ASPR has been pursuing a strategy to work with international partners to build a sustainable global infrastructure for medical countermeasures. For developed countries, one effort under way uses the Global Health Security Initiative (GHSI).[8] ASPR held GHSI medical countermeasure workshops in 2008 and 2009 to determine areas of interest for collaboration and to identify current gaps and barriers to international collaboration. GHSI will conduct an exercise to consider a single threat (anthrax) as a case study to identify gaps and concrete areas for collaboration. Major gaps and barriers to international collaboration identified included

- Countries perceive threats differently. There is a need for improved surveillance of threats, increased information sharing, and joint development of assessment tools.
- There is a need for information sharing to maximize resources and avoid duplication of efforts, for harmonized country regulatory requirements for market authorization and expedited clinical trial proc-

[8] GHSI is a forum for high-level discussion concerning the coordination of public health emergency preparedness and response policies for CBRN threats and pandemic influenza. It was launched in 2001 by the ministers of health of Canada, France, Germany, Italy, Japan, Mexico, the United Kingdom, the United States, and the European Commission. The World Health Organization serves as an expert advisor. A ministerial-level summit is held every year to share information and coordinate efforts to improve global health security. See http://www.ghsi.org/.

esses, and for innovative and modernized vaccine production processes.
- Countries should collaborate on point-of-care diagnostic tools, stockpiles, emergency deployment plans, and harmonization of treatments and use policies.

With regard to the developing world, Marinissen said that ASPR is initiating international discussions on a framework and strategic plan to create regional, independent, and sustainable influenza vaccine production capacity in developing and emerging economy countries. Such capabilities could then be used as a platform for surge capacity for pandemic vaccine.

Gillian Woollett, chief scientist, Engel & Novitt, LLP, cautioned that the United States might not want biopharmaceutical companies selling their countermeasures all around the world, and questioned the ability to control the use of an effective countermeasure. This could make the situation worse, she said, if the United States spent large amounts of money to develop a countermeasure, and someone buys it in order to protect his or her own people or engineers a different or resistant threat.

Issues for Federal Agencies Engaged in Countermeasures Development

Systemic Concerns

Philip Russell, Major General, U.S. Army (ret.), former senior advisor in HHS's Office of Public Health Preparedness and current member on the Board of Trustees of the Sabin Vaccine Institute, highlighted a number of systemic concerns impacting the effectiveness of the countermeasures enterprise. Reliance on an unwieldy and ineffective contracting process is a primary challenge across the board for all participants in the countermeasures enterprise. The Federal Acquisition Regulation (FAR) is unsuitable for product development in the pharmaceutical field, Russell said. The FAR restricts communication, and contractors generally lack experience and capability in the full range of skills needed to bring a medical product to licensure.

All product development paths must ultimately lead to the FDA, and the regulatory process can be cumbersome and fraught with uncertainty. Two key barriers were highlighted by workshop participants. The first, which will also be discussed later in this report, is the absence of a clear and consistently applied regulatory pathway for medical countermeasures. The Animal Rule is also presenting itself as a barrier. The Animal Rule says FDA may grant approval when "the results of those animal studies establish that the drug is reasonably likely to produce clinical benefit in humans." However, the guidance is significantly more restrictive than the Animal Rule itself and is being appropriately administered relative to the regulation of vaccines for biodefense, Mary Pendergast and Russell said.

As will be discussed later in the report, the lack of central leadership impacts the ability to bring together the numerous agencies involved in the countermeasures enterprise. (This topic was also a focus of a meeting hosted by the NBSB.) This viewpoint was shared by many workshop participants. However, others cautioned that although it is important to ensure that the DoD's efforts are aligned and coordinated, it may also be important to maintain a level of independence due to complementary, but separate, missions.

Project Bioshield and BARDA Resources

The Project BioShield Act became law in July 2004 (Public Law 108-276) and provides for procurement of countermeasures. However, many of the products it seeks to acquire are not yet available for purchase. As BARDA Director Robin Robinson explained, the 2006 Pandemic and All Hazards Preparedness Act (Public Law 109-417) included a corrective measure, establishing BARDA and provisions for supporting advanced development of products. BARDA is responsible for advancing projects funded by the NIH and the DoD, by moving them across the high risk advanced development zone (the "valley of death") to a point where acquisition and stockpiling can be achieved.

Although BARDA is off to an enormously good start, said Eric Rose, chief executive officer (CEO) and chair of Siga Technologies, it is a young organization on a steep learning curve. An economic analysis by Bradley Smith of the Center for Biosecurity, UPMC, and colleagues found that the advanced development mission of BARDA is underfunded by at least 10-fold and consequently its portfolio is very thin (Matheny, 2008).

As part of the Project BioShield Act, money for countermeasures procurement was set aside in escrow so that purchases could be made without needing to go back to Congress for an appropriation. Procurement authority for Project BioShield acquisitions using the Special Reserve Fund rests with BARDA. Chuck Ludlam, former counsel to Senator Joseph Lieberman and former principal lobbyist for the Biotechnology Industry Organization, and other participants expressed concern that money from the Special Reserve Fund is being diverted to other initiatives. For example, in the 2009 Omnibus Appropriations Act, $412 million was transferred to other programs to support countermeasure advanced research and development and pandemic influenza preparedness and response. In FY2010 an additional $305 million has been proposed to be transferred to support countermeasure advanced research and development. This could have serious repercussions on the government's ability to guarantee a suitable marketplace for future countermeasure procurement. This also leads to continued uncertainty among the private sector about long-term, stable funding for countermeasures research and development.

FDA Funding and Scientific Infrastructure

To highlight the challenges facing FDA today, Gail Cassell, workshop chair and vice president, Scientific Affairs, at Eli Lilly and Company reviewed the findings of the report *FDA Science and Mission at Risk* (FDA, 2007). The FDA Science Board Subcommittee on Science and Technology, chaired by Cassell, was charged by then-Commissioner Andrew von Eschenbach to review science and technology across the agency to answer the question of whether FDA is prepared to address emerging technologies in science. The subcommittee concluded that "science at the FDA is in a precarious position: the [a]gency suffers from serious scientific deficiencies and is not positioned to meet current or emerging regulatory responsibilities." Major findings of the report are presented in Box 4. The Subcommittee found that the deficiencies had two main two sources:

- The demands on the FDA have soared due to the extraordinary advance of scientific discoveries, the complexity of the new products and claims submitted to FDA for premarket review and approval, the emergence of challenging safety problems, and the globalization of the industries that FDA regulates.
- The resources have not increased in proportion to the demands. The result is that the scientific demands on the agency far exceed its capacity to respond. This imbalance is imposing a significant risk to the integrity of the food, drug, and device regulatory system, and hence the safety of the public. This also raises the issues of the threats associated with a bioterror attack and the critical role of FDA in the development of medical countermeasures.

BOX 4

***FDA Science and Mission at Risk*: Major Findings**

- The Food and Drug Administration's (FDA's) scientific base has eroded and its scientific organizational structure is weak.
- FDA's scientific workforce does not have sufficient capacity and capability (recruitment and retention challenges, insufficient investment in professional development).
- FDA is unable to keep up with scientific advances (systems biology, wireless healthcare devices, nanotechnology, medical imaging, robotics, cell- and tissue-based products, regenerative medicine, and combination products).
- FDA cannot fulfill its surveillance mission because it does not have adequate staff and IT resources to implement cutting-edge approaches to modeling, risk assessment, and data analysis.

SOURCE: FDA (2007).

Cassell noted that FDA had been given more than 100 unfunded mandates over the previous 15 years, while staffing did not increase concurrently to meet these new mandates. The agency is also responsible for conducting inspections at more than 300,000 sites in 100 countries. The FDA has a huge economic impact, regulating 25 cents of every dollar that Americans spend—over $1 trillion worth of products ranging from cosmetics to pet food. Yet in 2007, FDA had an appropriated budget of only $1.6 billion, which is about 1.5 cents per day per American.

While the agency has made progress in addressing each of the major deficiencies noted in the report, much more needs to be done because regulatory and information sciences are the very foundation of the FDA's mission. They are critical to the agency's role in development of medical countermeasures for biodefense. Although the world of drug discovery and development has undergone revolutionary change—shifting from cellular to molecular and gene-based approaches—FDA evaluation methods have remained largely unchanged over the past half century. Likewise, evaluation methods have not kept pace with major advances in medical devices and use of products in combination.

The Subcommittee noted that the impact of the deficiency is profound precisely because science is at the heart of everything FDA does. The world looks to FDA as a leader—to integrate emerging understandings of biology with medicine, technology, and computational mathematics in ways that will lead to successful disease therapies.

Today, not only can the agency not lead, it cannot even keep up with the advances in science. Due to constrained resources and lack of adequate staff, FDA is engaged in reactive regulatory priority setting or a firefighting regulatory posture instead of pursuing a culture of proactive regulatory science.

The Subcommittee identified the following eight emerging science and technologies that are the most challenging to the FDA: systems biology (including genomics and other "omics"), wireless healthcare devices, nanotechnology, medical imaging, robotics, cell- and tissue-based products, regenerative medicine, and combination products. Each of these emerging areas is developing at an exponential rate and each generates novel scientific, analytic, laboratory, and/or information requirements. These areas are also precisely those that have been identified as being critical to development of medical countermeasures. Furthermore, the FDA cannot fulfill its surveillance mission because of inadequate staff and IT resources to implement cutting-edge approaches to modeling, risk assessment, and data analysis. The status of regulatory and information sciences at FDA must consider our ability to successfully address the threats of bioterrorism. Other participants concurred, noting that there is no surge capacity at the FDA, or that in fact it is already operating at surge capacity.

Challenges Facing the Innovative Biopharmaceutical Industry

A focus of the workshop was to identify how to improve innovation in ways that respond to national priorities, including how to better engage the nation's commercial drug, biologic, and device manufacturers in the countermeasures enterprise. Participants from industry described a variety of barriers and challenges to commercial involvement.

Risk and Uncertainty

Most new or in-development pharmaceutical products fail, said John Rex, vice president and medical director for infection at AstraZeneca. Philosophically, a company starts with that understanding, and designs programs to manage risk and to identify failures quickly and cheaply, without committing too many resources, until there is a reasonable level of confidence in the product. The pharmaceutical industry is very good at models and methods to help address the scientific, technical, formulation, and safety risks, for example. The risks that drive industry away occur when changes happen that cannot be readily anticipated—for example, when regulatory guidance is not clear.

Regulatory uncertainty at the FDA was a recurring theme during the discussions. For small companies in particular, this uncertainty can be compounded by a limited understanding of the regulatory pathway. The investment community also has a keen interest in the success or failure of industry pursuits. A participant from an investment bank, Stephen Brozak of WBB Securities, LLC, noted that "Wall Street hates uncertainty." FDA adds uncertainty because financial analysts have no way to quantify how long the approval process will be for a product. The new draft guidance from the FDA caused significant uncertainty that hampered investors' ability to predict regulatory, and consequently, revenue trends. If the FDA could establish a clear regulatory pathway in biodefense, it would allow the analysts some sort of metric to be able to say "if a company does this, that will happen." That is likely to foster greater interest and investment in companies doing research in countermeasures, a point highlighted by multiple workshop participants.

Material threat *determinations* (the list of pathogens of concern) are public, but material threat *assessments* and population threat assessments are classified. So while the PHEMCE does provide some highly desirable predictability with regard to identification of the targets for discovery and development, companies have no information regarding the planning scenario for which they are trying to build a product. Although it is understandable that industry is not included in the PHEMCE, this leaves the countermeasures enterprise itself with a critical lack of business and capital markets expertise. The PHEMCE implementation plan itself provides limited guidance and is essentially a list of pathogens and agents that the government hopes to acquire. This kind of checkbox approach obscures product shortcomings and regulatory gaps. Participants also suggested that the implementation plan is somewhat counterproductive in that it defines the market, arbitrarily, as either above or below $100 million. Consequently this means that companies will not invest in developing products predicted to gross less than $100 million.

Not knowing how a product is going to be commercialized is also a risk that industry prefers to avoid. The manufacturing of biologics is complex, and there is an enormous difference in manufacturing, for example, 200,000 doses versus 40 million doses. Companies need guidance regarding volume so they can develop manufacturing plans.

The acquisition process (for initial stockpiles until product licensure), which is essentially guided by the Federal Acquisition Regulation, is perceived by the industry to be lengthy, opaque, unpredictable. In particular, the transition trigger from advanced development to acquisition Request For Proposals (RFP) is unclear. After acquisition, there is a perceived improved communication compared to acquisition process, but some aspects remain unclear, particularly FDA coordination with BARDA. As a result, Goodman's key principle of end-to-end partnering, including highly interactive and collaborative engagement and outcomes-oriented management, takes on an increasingly important role.

Financial and Resource Concerns

The most often mentioned financial barrier to engaging biopharmaceutical companies in the countermeasures enterprise is lack of market incentive. Wesley Yin, assistant professor in the Department of Economics at Boston University, said that firms simply are not going to be able to recoup the fixed costs of research and development of countermeasures. Unlike a standard low-prevalence disease, not only is demand low, but it is also uncertain. If there is demand, it usually comes in times of public health emergency. Plus, there is pressure, real or perceived, to sell these technologies at or just above marginal costs. In addition, if a company overcomes the revenue risks and pursues development of a product, it is at risk for product liability issues, which are also a financial and resource burden.

Lack of market incentive aside, Thomas Monath of Kleiner Perkins Caufield & Byers pointed out, participation in the countermeasures enterprise has a huge opportunity cost—taking away a company's ability to focus on its commercial opportunity market. Woollett, of Engel & Novitt, added that the industry is actively making products that are saving lives, and asked whether those lives are any less important than a putative potential threat. Simply adding a capability is not an option unless we are prepared to take away from something else, she said.

Participants also highlighted that while companies are interested in countermeasures development, the long-term financing piece must be addressed to be able to make a more rational business case for devoting company resources to countermeasures. The ability to plan for the future can speed up everything tremendously (e.g., if the first step is successful, the company can move to the next step, and already be planning for the next clinical trial, without an interim funding step).

An issue for biotechnology companies is that they are generally small, unprofitable entities that are sustained by private capital and government grants and contracts. Many do not survive. But these companies are an integral part of the PHEMCE implementation plan. Rose of Siga Technologies said BARDA has been an excellent, responsive development partner for the biotechnology industry. BARDA funding for direct project costs are reasonable, realistic, and flexible. He said, however, that the funding for *indirect costs* is only a fraction of actual costs, and has to be supplemented by in-house funding and private capital in order to keep these projects going for the 8 to 12 years that drug development generally takes. For small innovative companies, programs may start and stop, but funding cannot be discontinued and started again. Small organizations depend on that ongoing revenue to continue to employ staff.

Another enormous drain on resources for small biotechnology companies is the RFP process, said David Wurtman, vice president at NexBio. Although it can be quite constructive for companies to think through the entirety of a development-to-manufacturing plan, if the company is small, research may come to a halt as all hands focus on the RFP. Despite the efforts companies make to respond to an RFP, they often do not find out if they have been awarded a contract, which can present difficulties in planning for the future, especially if the company is small. From a human resources perspective, advanced development manufacturing is really an apprentice model. University training to grow a pool of talent is limited, if it even exists, said Phillip Gomez of PRTM. Therefore, it is important to provide opportunities for partnerships between academia and industry, where the advanced development manufacturing expertise rests. This will help grow the base of people with this expertise and help people learn from a variety of perspectives in the enterprise.

Intellectual Property and FDA Approval

The protection of intellectual property is at the core of the industry's ability to earn a return on research investments and remain competitive. The ideal situation, according to Bruce Artim of Eli Lilly and Company, would be for an innovator to be awarded a patent by the Patent and Trademark Office (PTO) on the same day FDA approval is granted. However, in practice, FDA product approval review generally takes much longer than PTO patent application review, effectively reducing the patent protection period. A significant policy challenge is balancing two needs: (1) the need of the innovator drug company both to recoup costs and to profit and grow so it can continue to innovate, and (2) the need to bring less expensive generic versions of products to market. The 1984 Drug Price Competition Patent Term Restoration Act, commonly referred to as the Hatch-Waxman Law, is extremely complex, but at its core, it allows a generic drug manufacturer to refer to the pioneer drug developer's data when applying for FDA approval of the generic form.[9] The pioneer also receives additional years of patent term to compensate for some of the time the drug was already on patent while still under the lengthy clinical development and FDA review periods. The pioneer also receives 5 years of data exclusivity. During that time, FDA cannot approve any generic drug applications for a comparable product. Basically, Artim said, this incentive system places more importance on patent term as an intellectual property tool than data protection. Therefore, companies invest resources where they believe they have strong patent protection. However, no correlation exists between the patent protection and the scientific or clinical value of the molecule.

Special Considerations for Antibiotics

Because small-molecule antibiotics have dual uses (both as standard medical care and as medical countermeasures), one might think they would be the countermeasure with the simplest development pathway. But this is not necessarily the case, said Rex of AstraZeneca.

A variety of considerations are specific to the development of antibiotics. First, Rex said, discovery and development are iterative. Simple "gateway indications" provide the entry point, and securing approval for the gateway indication (e.g., community-acquired pneumonia) opens the door to many other uses down the road, including countermeasures. But if a company cannot achieve approval for the basic clinical indication, nothing will follow. Second, bacterial resistance drives the need for novel antibiotics. The ideal comparative clinical trials—new drug versus the drug to which the organism is resistant, or a placebo-controlled superiority study—simply cannot be done for obvious ethical reasons. Rather, non-inferiority designs versus an active agent must be used. This approach has caused significant regulatory confusion. Non-inferiority trial design is more difficult to implement than superiority designs. Following approval, the new drug is subsequently perceived as only non-inferior rather than superior because its activity when other drugs would be resistant is not apparent. Finally, there is the paradox of antibiotic value. A new antibiotic may be deemed so important that it is not used, reserved only for situations when all else fails, which presents a problem for companies who plan to recover some of their development costs though sales. Pricing of the new antibiotic is also a challenge, especially when the new drug has only been shown to

---

[9] In filing an Abbreviated New Drug Application, the generic manufacturer does not have to conduct clinical trials; rather, it must demonstrate that the generic product is bio-equivalent to the innovator drug.

be non-inferior to an existing generic drug, raising the question of why it should then cost more.

Research Infrastructure and Resources

Tools and Methodology

Infrastructure for rapid countermeasure development requires common data elements in both research and practice across the board, summarized Marietta Anthony from the Critical Path Institute. A unique scientific issue for countermeasures research is not having the disease to study in many cases, and the development of clinical disease/clinical injury models for trial simulation would be very useful to help speed the process. Innovation also needs to be brought to clinical trial design (e.g., adaptive clinical trial design). Biomarkers that are qualified for use by the FDA to reliably and accurately detect diseases in the field, or detect changes in the field, are also needed. Rapid point-of-care testing and resistance testing was cited as a need by state health departments.

Whether products are to be for engineered threats or natural pathogens, other research needs include vaccine adjuvants, cell culture manufacturing, expansion of biologics manufacturing capacity, and decontamination and remediation protocols after an attack or exposure.

Academia

In general, academic research, and to a large extent government research, are not intended to produce products. In academia, grant funding and publications are highly valued and are the currency for tenure or promotion, noted Brett Giroir, vice chancellor for research at Texas A&M. Product development, intellectual property, and commercialization, while not discouraged, are generally not fostered or rewarded.

The basic research funded by government and conducted in academia is, in general, not prioritized by national need. If research is successful in identifying a potential product, there are no transition partners lined up and no clear pathways for investigators to carry their discovery forward. As a result, it is likely that government and industry are aware of only a very small fraction of the innovations from academic laboratories that could eventually lead to products, Giroir said.

Liability

While believing that the 2006 Public Readiness and Emergency Preparedness (PREP) Act (Public Law 109-148)[10] went a long way to address liability issues, a number of participants cautioned that there are gaps and holes that have not been filled. For example, it is unclear if PREP Act declarations preempt state tort law, which can result in continued liability concerns for end users. There is also liability in terms of what must be disclosed for informed consent in case of an emergency.

End Users: Challenges for Public Health and Providers

Although public health officials at the state and local levels are not directly involved in the research and development of countermeasures, they are responsible for ensuring the safety of the public by implementing whatever comes out of the countermeasures enterprise. Therefore, the needs of the public health as end users should inform the target product profiles. State and local public health and healthcare providers all play a critical role in the delivery of countermeasures. The need to integrate these individuals much earlier into the process of research and development of the countermeasures, perhaps through an advisory board to BARDA, was highlighted at the workshop as an opportunity.

Susan Cooper, commissioner of the Tennessee Department of Health, commented that although an abundance of product variations may seem like a benefit, it adds significant complexity to state implementation activities. For example, influenza vaccines come in single-dose syringes, multidose vials, and intranasal mists, each with its own labeled uses in different subpopulations. These different products are shipped as they became available, making distribution to different real-time providers a challenge. She also noted that the variety of forms of the vaccine confounds public health messaging. The Advisory Committee on Immunization Practices guidelines, for example, identified the priority groups to be vaccinated first as pregnant women, children with chronic diseases, healthcare workers, persons between the ages of 6

---

[10] A "PREP Act declaration" by the HHS Secretary provides immunity from tort liability (except for willful misconduct) for claims of loss associated with the administration or use of medical countermeasures to threats that are deemed by the Secretary to constitute a public health emergency, to those involved in the development, manufacture, testing, distribution, administration, and use of such countermeasures (http://www.hhs.gov/disasters/discussion/planners/prepact/index.html).

months and 24 years, and persons from ages 25 through 64 years who are at higher risk for novel H1N1 because of chronic health disorders or compromised immune systems. Unfortunately, following statewide media campaigns urging these groups to be vaccinated, the first product received in Tennessee was the intranasal mist—which cannot be given to pregnant women, children with chronic disease, and those over the age of 50, which would include many healthcare workers. So although states understand the challenges of developing countermeasures and acknowledge that choice is important, an abundance of choices can actually complicate implementation.

Although not a focus of this workshop, a recurring theme was the importance of investing in the public health infrastructure and delivery, noting that the most effective products have no value if you cannot get them to people who need them, or if people do not trust the product. State and local public health departments will require epidemiologic and laboratory capacity; robust emergency drill programs; management, logistics, and communication capacity; strong links with healthcare systems; and integration across a variety of non-health sectors (e.g., police, transportation, education). Participants specifically called out rapid point-of-care testing and resistance testing as current needs, noting that the ability to detect resistance in anything close to real time is very difficult.

EXAMPLES OF SUCCESSFUL COUNTERMEASURES DEVELOPMENT AND DEPLOYMENT

Features of Successful Government Countermeasures Efforts

Russell, identified earlier, presented several examples of products that were successfully developed by the government over the past 30 years, including the U.S. Army's development of products for adenovirus, meningococcus, hepatitis A, nerve gas, and malaria, and recent HHS efforts on smallpox. Looking across these successful programs, Russell noted that several shared characteristics emerge, including good scientific direction and leadership, a strong pharmaceutical manufacturer as a partner, and the internal capability to move the candidate through pilot level (Box 5).

Captain Kenneth Cole, medical director of the Nuclear and Chemical and Biological Defense Programs at DoD, said a key element of the DoD approach to countermeasures research is an *oversight mechanism* that considers the entire portfolio. Within the DoD there is the Joint Science and Technology Office, which handles basic research and early development; the Joint Program Executive Office for Chemical and Biological Defense, which handles advanced development and procurement; and the Joint Requirements Office, which is involved in planning, coordination, and oversight; defining desired capabilities; defining requirements needed to get there; setting key performance parameters; and determining how the product is going to be used. Within that program is a *single authority* that can make the decision as to when something translates from research and early development into advanced development. As part of the oversight process, a Medical Advisory Board looks at the full spectrum of research and advance development, and makes recommendations to this milestone decision authority on which candidates to advance.

This structure allows for oversight, accountability, prioritization, and translation of requirements all the way to the research level. Throughout the process, DoD leverages interagency as well as international partners (through 64 different bilateral, trilateral, and quadrilateral chem/biodefense relationships and treaties around the world).

BOX 5
Characteristics of Successful Government Product Development Programs

- Direction and management by scientists experienced in the product development process, and empowered program managers in the government
- Close working relationship with a major manufacturer/partner, with open, effective, and direct communication
- Good leadership from both government and industry
- In-house capability for process development, pilot manufacturing, and clinical trials
- Full support of senior leadership and recognition of the need
- Good working relationships with the FDA
- Minimal interference by the contracting officers in the process and the communication between the government developer and the industrial developer
- Little or no interference by Congress or lobbyists

Transformational Medical Technologies Initiative

A new DoD initiative is the Transformational Medical Technologies Initiative (TMTI), specifically designed to look at the emerging and bioengineered threats the warfighter faces, the DoD's Cole continued.

The goal is to move from the traditional "one bug–one drug" approach to the transformational "one drug–many bugs" approach, developing broad-spectrum countermeasures (e.g., that target common disease pathways or enhance the host's immune response), as well as platform technologies to characterize unknown pathogens and rapidly develop medical countermeasures to newly identified threats. The TMTI approach integrates efforts within government, academia, the biotechnology industry, and small and large pharmaceutical corporations, providing seamless "end-to-end" product development.

Cole suggested that a national medical countermeasures strategy needs to be structured with an oversight and accountability mechanism that drives the requirements from research through advanced development and provides a focus to the program. In this regard, TMTI is an excellent model.

Transformational Medical Technologies Initiative: Demonstrated Capability

A rapid response biodefense capability is the overarching TMTI goal, including broad spectrum medical countermeasures and platform technologies. As a response to needs outlined in the 2006 Quadrennial Defense Review, since 2007 TMTI has seen two Investigational New Drug (IND) submissions submitted for hemorrhagic fever viruses and is preparing two broad spectrum antibiotic INDs, three broad spectrum antivirals, and is establishing efficacy for two innate immune activators. TMTI is also repurposing drugs already licensed for use as broad spectrum antibiotics. There are over 20 candidates in the pipeline. Most importantly, TMTI has demonstrated a capability to respond to emerging threats by producing an antiviral to the emerging H1N1 virus within seven days of receiving tissue samples and then proving efficacy in standard mice and ferret models. TMTI is so successful, the Department of Defense is transitioning it from an initiative to a program of record. The TMTI strategy is to move hemorrhagic fever viruses therapeutics through IND into the regulatory critical path, move intracellular bacterial pathogen therapeutics through IND submission into the regulatory critical path, advance animal models suitable for pivotal animal studies supporting licensure, develop pathogen gene lists and attributes for assessing pathways and target identification in order to recommend medical countermeasures, and finally to integrate all of these activities to identify candidate medical countermeasures against unknown pathogens. To date, hemorrhagic fever viruses candidates are moving onto clinical trials, intracellular bacterial pathogen candidates are moving into clinical trials, animal models continue through development and validation, end-to-end pathogen evaluation and medical countermeasure development has been demonstrated, and discovery along with pre-clinical efficacy, safety, and toxicity studies continue on further candidates for pipeline replenishment.

Lessons from Pandemic 2009 H1N1 Influenza

Goodman of the FDA also cited the recent response to pandemic 2009 H1N1 influenza as an example of success. A great deal of investments were made and significant planning was done, including at FDA, and there were excellent public–private partnerships and inter-agency collaboration and communication. One important aspect, he said, was that FDA set up an Incident Command System (ICS) to address the urgent needs presented by 2009 H1N1, which was extremely valuable. FDA staff were able to collaborate with other federal agencies and rapidly respond in an ICS mode.

Daniel Jernigan, deputy influenza director for the CDC, described seven actions that serve as the strategy for diagnostic preparedness. They are as follows:

1. Developing new diagnostic tests and improved capabilities
2. Improving surge capacity
3. Implementing proficiency testing
4. Developing policy and regulatory preparedness to facilitate rapid responses
5. Improving access to viruses and reagents
6. Providing guidance for clinicians
7. Improving overall virologic surveillance

Another key feature of a successful program is adaptability, said Andrew Pavia of the University of Utah School of Medicine. Planning only works to a certain point. Problem solving and flexible approaches need to be integrated into countermeasure development and emergency planning. For example, because of a study that was funded by NIAID through the Collaborative Antiviral Study Group, there were data that had not yet been published on the appropriate dosing of Tamiflu® for children with influenza. Because of a flexible response by FDA and CDC, those unpublished data were able to be used in the EUA and providers were able to treat children and save lives.

Countermeasures Development in Industry

Focusing on Unmet Medical Need, Deriving Dual-Use Products

In response to the Graham-Talent Commission report on the *Prevention of Weapons of Mass Destruction Proliferation and Terrorism* described earlier, the Center for Arms Control and Non-Proliferation said in a statement published on January 26, 2010, "Direct targeting of effort and expenditure on natural disease threats would provide much greater public health benefit, and spinoffs from these programs would significantly strengthen resistance to bioterrorism" (Center for Arms Control and Non-Proliferation, 2010).

George Painter, CEO of Chimerix, said his company follows the strategy outlined by the Center for Arms Control and Non-Proliferation. The company is structured to focus on unmet medical need, and if there is an opportunity to expand from that unmet medical need into a countermeasure area, the company will pursue it. For example, Chimerix was actively developing an orally available broad-spectrum antiviral drug to treat double-stranded DNA viral infections in immuno-compromised patients. Orthopoxviruses (which include *Variola*, the agent of smallpox) are also double-stranded DNA viruses, and the company bridged into biodefense on a grant from the NIAID. Similarly, the company has a robust hepatitis C drug development program. Hepatitis C virus and dengue virus are both flaviviruses, and the company will leverage what is learned as it takes the hepatitis C lead candidate into clinical development to help address dengue.

By managing the portfolio in this manner, the company has a clear regulatory pathway to approval and a definitive, definable market size that engenders interest on the part of private capital. The company can leverage that position if the opportunity arises to maintain a position in biodefense. Another advantage of pursuing only drugs with multiuse potential, Painter said, is that a clinical treatment protocol can be left open, and in the case of an emergency, one can broadly treat through that protocol.

Tyler Martin, chief medical officer of Dynavax, said that like Chimerix, Dynavax has focused its development efforts on commercial targets and then looks to see what other value can extracted from the research discoveries already made. Dynavax is a small biotechnology company of about 100 people, focused on exploiting the biology of Toll-like receptors for product development. "Portfolio management," he said, "is something that a biotechnology company does every day, trying to create maximum value from a limited number of resources." The way to do that is with target product profiles. Martin described three dimensions of the profile that are malleable: Change the scope of the assignment (increase or decrease the target product profile), change the resources available for the assignment (money, people, etc.), or change the time line to complete the assignment.

Melinda Moree of BIO Ventures for Global Health noted that the United States is not necessarily focused on products for the developing world, but innovations for these regions may have a dual application for countermeasures development. For example, a heat-stable vaccine that does not require refrigeration and has a long shelf life would be more appropriate for a stockpile than a product with a shelf life of 18 months. David Gilbert of Providence Health & Services commented that efforts to develop new antibacterials for multidrug-resistant, gram-negative rods for the civilian population would benefit the countermeasures enterprise as well.

Countermeasures as Orphan Drugs

An orphan disease in the United States is one affecting fewer than 200,000 people. Marlene Haffner of Haffner Associates and former director of the Office of Orphan Products Development said that since passage of the Orphan Drug Act in 1983, 18 products have been designated as potential countermeasures for terrorism, 4 of which have been approved (for exposure to cyanide, cesium or thallium, and pediatric exposure to radioactive iodine).

Although the orphan drug program has been able to support development of products in the arena of counterterrorism, it has drawbacks. The Orphan Drug Act gives 7 years of exclusive marketing of that product for that indication, which means that another company cannot approach FDA with the same product for 7 years. This could be a concern if a goal of preparedness is to have redundancy. The other main incentive of the Act is tax credits for clinical trial expenditures. But because clinical trials may not be possible in some cases of countermeasures development, tax credits may not always serve as a compelling incentive.

PARTNERSHIPS AND ALTERNATIVE BUSINESS MODELS

Venture Philanthropy and Orphan Product Development Models

As discussed above, there are several examples of successful medical countermeasures development under the Orphan Drug Act. Nonprofit disease research organizations and venture philanthropy groups are a force behind much of the progress in orphan product development. These organizations were primarily founded by patients because there was not enough research focus on their particular disease area (IOM, 2009). Their model is to derisk the research. Many of these groups are partnering successfully with industry, approaching biotechnology companies directly and offering funding for research in their area of interest. Margaret Anderson of FasterCures described two new rare disease-related activities that, if implemented effectively, may also serve as an opportunity for improved medical countermeasure development—the Cures Acceleration Network (CAN) and the Therapeutics for Rare and Neglected Diseases (TRND) program.

The Cures Acceleration Network Championed by Senator Arlen Specter and part of the recently passed Patient Protection and Affordable Care Act (Public Law 111-148) CAN seeks to cut the time between discovery and development of drugs and therapies through new grant-making mechanisms at the NIH; establishes CAN within the Office of the Director of NIH and authorizes grants to move discovery from the lab into the next generation of therapies; and integrates the FDA into the work that CAN will undertake, providing a vital link between NIH and the drug approval process.

Therapeutics for Rare and Neglected Diseases program at NIH TRND is a congressionally mandated effort to encourage and speed the development of new drugs for rare and neglected diseases. Specifically intended to stimulate research collaborations with academic scientists working on rare illnesses, TRND supports specific, preclinical research and product development, leveraging the in-house scientific capabilities needed to carry out much of the preclinical development work and contracting out other parts, as scientific opportunities dictate (http://rarediseases.info.nih.gov/TRND).

Pharmaceutical Shared-Risk Approaches

The private sector is also in the process of a significant reorganization, where it is establishing a large number of partnerships—public–private and private–private—that can leverage the expertise of multiple sectors, thus reducing costs, increasing probability of success, and sharing risk. For example, Eli Lilly has moved away from a "fully integrated pharmaceutical company" model, where everything is done (and correspondingly, funded) internally, to a "fully integrated pharmaceutical network" model, explained Paul Owens of Lilly

Research Laboratories. Lilly is looking to access global networks in a virtual fashion and thereby be more productive, optimizing the speed and cost of drug development. Companies can no longer afford to have the source of ideas, the source of capacity or capability, and the source of funding all under one roof, he said.

The **Phenotypic Drug Discovery** program is one example of the fully integrated pharmaceutical network approach. Lilly has opened up phenotypic screening models in five disease areas of interest (Alzheimer's disease, osteoporosis, diabetes, cancer cell growth, and cancer antiangiogenesis) to any interested external drug discovery entities (e.g., academic institutions, small biotechnology companies, individual faculty). Investigators can submit a molecule into these phenotypic drug screening panels in a confidential manner. If there is a hit, a secondary, more target-based screening will be conducted. After the full set of five panels is completed, a report is sent to the investigator. The investigator retains the intellectual property, but Lilly has the first right of negotiation for a commercial opportunity on that particular hit for a defined period of time after the screen is completed. This program has attracted hundreds of samples from all over the world. A platform of this type could be applied to many different disease areas (http://www.PD2.lilly.com).

Another example is the **TPG-Axon/NovaQuest collaboration with Lilly.** TPG-Axon is a venture capital firm and NovaQuest is the investment arm of Quintiles, a global clinical research organization. Together, this three-way partnership will share the risk , and the potential return on investment, for the development of Lilly's two lead candidates for the treatment of Alzheimer's disease. Although Lilly is giving up some value of these assets, disseminating some of the risk allows the company to invest in a larger portfolio all the way back into discovery.

Planning for Failure

As described above, several industry participants described how failure is an inherent risk in biopharmaceutical development. The sign of a mature discovery group, said Rex of AstraZeneca, is the ability and willingness to stop pursuing drug targets or candidates. Scientists can become very attached to their targets, and can be unwilling to move on regardless of lack of progress. In addition, people are less willing to abandon the only targets they may have if they perceive their employment or the survival of their project is dependent on finding a candidate from a small pool of opportunities. To address this,

AstraZeneca has removed the process of hit-and-lead candidate identification from the responsibilities of individual biochemists. Instead, a small group is now charged with assessing leads for candidates that might be drugs. There is no emotional or historical attachment to a particular target or molecule, just a focus on identifying drug leads with good, drug-like properties.

Similar to the AstraZeneca approach, Eli Lilly accepts that failures will happen and looks to identify them early, commented Owens. **Chorus** is a small autonomous drug development group within Lilly focused on moving molecules from candidate selection through to proof of concept as quickly and inexpensively as possible. The goal is to eliminate as much risk as possible—pushing a molecule to a point where it works or not—and facilitating the decision of whether to invest in Phase III clinical trials or not.

Along the same lines, Lilly is now starting to develop a new **lead-optimization program,** a small, flexible group of scientists who are autonomous and independent of the larger Lilly Research Laboratories. They will take potential hits from the Phenotypic Drug Discovery program or another end-license opportunity, and pull together the required data elements to make decisions regarding candidate selection.

In addition, Owens noted that Lilly essentially suspends scale-up manufacturing and all major chemistry, manufacturing, control studies and formulation strategies until post-proof of concept, eliminating both time and cost.

Open Innovation Business Strategies

Open innovation is very important for the development of breakthrough medicines, said Teri Melese, director of Research Technologies and Alliances, University of California–San Francisco School of Medicine. More effective classification is needed from companies about what data and information can and cannot be shared (Melese et al., 2009). Melese cited several models of open innovation, which may serve as valuable models for the countermeasures enterprise (Box 6).

BOX 6
Models of Open Innovation Business Structures

Novartis/Broad Institute Diabetes Genetics Initiative: The initiative will perform whole-genome scans on DNA collected from type 2 diabetic patients worldwide to provide a comprehensive view of the DNA sequence variants associated with the disease. Genome data will be made publicly available.

GlaxoSmithKline (GSK) Open Laboratory for Tropical Diseases: Up to 60 scientists from around the world will pursue their own projects as part of an integrated team and will be afforded access to the Open Lab based in Spain, as well as to the expertise, knowledge, and infrastructure of the company. GSK will also make its library of 13,500 malaria compounds publicly available.

Lilly, Merck, and Pfizer Asian Cancer Research Group, Inc.: This is an independent, not-for-profit company established to accelerate research and ultimately improve treatment for patients affected with the most commonly diagnosed cancers in Asia. In this precompetitive collaboration, the companies will combine their resources and expertise to advance knowledge of disease and disease processes.

The Merck Gene Index: This is a collaborative effort begun in 1994 to release vast human genome sequence information from Merck's gene index into the GenBank and the public domain.

Public–Private Partnerships

Public–private partnerships were of special interest to many participants representing all sectors of the countermeasures enterprise—small biotechs, large pharmaceutical companies, academia, and government. While industry involvement in medical countermeasure product research and development is important, it cannot be relied upon to solve all the problems. Large pharmaceutical companies have too many disincentives and will not enter this arena sufficiently to get the job done. The small biotechs, while likely to come up with some novel development candidates, lack the experience and capacity to take a development candidate through all the clinical testing and regulatory hurdles, and they lack manufacturing capacity. Public–private partnerships will also be valuable for improving access to the chemical libraries and chemistry expertise of the larger companies to develop antiviral and antibacterials in the preclinical phase and early discovery, something that likely cannot be accomplished alone through grant mechanisms. By leveraging the specific expertise of each of these sectors the entire countermeasures enterprise would be advanced, commented

Michael Goldblatt, president and CEO of Functional Genomics. In addition, the establishment of these partnerships would help entice further investment by industry. A variety of models were presented. Generally, industry supplies a list of priority areas of interest and problems they want to solve. For their role, the industry partner(s) supplies scientific advice and access to commercial resources (e.g., compound libraries, medicinal chemistry, manufacturing knowledge) to support the consortium. In return, the industry partner generally wants first rights to commercialize any resulting intellectual property.

One relevant, successful government-initiated public-private partnership is the **National Space Biological Research Institute,** which is a NASA-funded academic consortium to procure countermeasures for health-related issues associated with space flight. Industry members participate as partners, advisors, collaborators, and consultants, helping to accelerate product development (http://www.nsbri.org/).

Another example is the recently established **Merck/Wellcome Trust Hilleman Laboratories,** the result of a charity and a company investing equally to form a new entity. The nonprofit Hilleman Laboratories will leverage scientific expertise and platform technologies for discovery development of vaccines for the developing world.

Examples of product development partnerships that the Bill & Melinda Gates Foundation, the U.S. government, and others have funded to advance the development of vaccines, drugs, and diagnostics for poor populations around the world include the **Malaria Vaccine Initiative,** the **Human Hookworm Vaccine Initiative,** and the **Global Alliance for TB Drugs.** As a result of those investments, Rajeev Venkayya of the Bill & Melinda Gates Foundation said, 86 different products are in the pipeline at various stages of development (the majority in preclinical development and Phase I clinical trials). One aspect of these successful models that has been valuable, Venkayya said, is that the Foundation has offloaded the coordination of these activities. A benefit of product development partnerships, such as the Malaria Vaccine Initiative, is that there is a portfolio of products against one target. This facilitates learning across projects, and the ability to make go/no-go decisions based on an entire portfolio of products against one target.

Having looked at many models of success and failure, Russell, identified earlier, recommended public–private partnership as a critical aspect of product development success, and listed several key elements of such partnerships for countermeasures development (Box 7).

BOX 7
Elements of Public–Private Partnerships for Countermeasure Development

- Fully dedicated to government requirements
- Funded by government with cofunding from local organizations
- Oversight by government agency
- Managed by a nonprofit organization
- Operated by expert personnel
- Industrial partners include vaccine and equipment companies
- Multiple options for working with innovators
- Multisuite flexible manufacturing facility
- State-of-the-art manufacturing technology

Gerald Parker, principal deputy assistant secretary in ASPR, suggested that in public–private partnerships involving product development, commercial scale-up, and manufacturing, the establishment of *technical centers of excellence* could provide dedicated capabilities and experienced technical personnel, particularly for these low-probability/high-consequence medical countermeasures that are needed for national security and emergency public health.

Another model would be a U.S. government support organization. The purpose would be to help specifically advance single-use targets where large pharmaceutical companies or small biotechs could not be relied upon to take on such targets due to minimal profit margins. This entity could focus on research and development for such products. Due to the intersection of the public and private sectors, leadership would be an important consideration, with an ideal person having background in industry to oversee discovery/development of important drug/vaccine products. This entity should have access to all the latest technology through licensing, if necessary, and would develop product candidates through phase 2. Further product development could then take place through contract research and development organizations and manufacturing could be contracted to commercial organizations. However, government manufacturing centers could also be used, if there is inadequate manufacturing capacity available for such products (such as vaccines).

A question was raised as to whether such synergy between the public and private sectors would raise concerns of conflict of interest, particularly partnership involving FDA and industry. Anthony Fauci, director of the National Institute of Allergy and Infectious Diseases, said the focus is on getting the FDA resources to be able to become scientifically involved in aspects of development that can be used by any drug innovator (e.g., biomarkers, alternative clinical trial designs), so in that context there is no conflict. Goodman concurred, noting that working very iteratively and highly interactively with sponsors on trying to achieve an important public health outcome would not be a conflict of interest.

Independent Third-Party Facilitation of Collaboration

The Critical Path Institute was developed in response to the FDA's Critical Path Initiative (FDA, 2004), with the intent of establishing a neutral ground where multiple companies could come together to work with the regulatory agencies (FDA, European Medicines Agency), with scientific experts from the NIH and academia, and with patient advocacy organizations.

Anthony of the Critical Path Institute said the real purpose of this model is to foster collaboration both among companies and between those who are regulated and the regulators. Anthony noted that typically, the contents of an individual company's discussions with FDA are proprietary, and therefore any successes or failures remain confidential and other companies are subject to repeating those failures. This is an attempt to improve the process, where companies share data in the precompetitive area—not on products, but on tools and methods that can improve the process—with the goal of developing scientific consensus on tools and methods among those who are going to use them and those who will accept them (the regulators).

Anthony described three consortia that are models of collaboration: the **Predictive Safety Testing Consortium,** developing biomarkers for safety; the **Patient-Reported Outcomes Consortium,** developing, evaluating, and qualifying instruments for use in collecting information directly from the patient; and the **Coalition Against Major Diseases,** sharing information about basic mechanisms of disease.

Strategic Investor Model

Although every year \$30–\$65 billion in new investments from venture capital and industry support innovation and biotechnology, almost none of that money goes toward countermeasure development, commented Monath, of Kleiner Perkins Caufield & Byers (PhRMA, 2010). The government should act as a *strategic investor*, in partnership with private capital. In the case of countermeasures development, the government would invest in technology development in a company, and the return on investment would be products. Thus the government would be rewarded for sharing the risk.

There is a precedent in government for this strategic investment approach. The Central Intelligence Agency established a venture capital company, In-Q-Tel, that makes investments in leading-edge commercial technologies that can help enable the agency's mission and broadly serve national security interests (Box 8).

BOX 8
In-Q-Tel

- Chartered by the Central Intelligence Agency in 1999 as an independent, nonprofit entity, now serving seven U.S. government agencies and open to working with additional agencies
- Engages a range of companies from start-ups to established companies, as well as universities, research laboratories, entrepreneurs, and venture capitalists
- Development agreements are aimed at building capabilities for the In-Q-Tel client base
- Takes equity in the companies in which it invests
- Concentrates on three broad commercial technology areas: Physical and Biological Technologies, Security, Software and Infrastructure

ENGAGING INDUSTRY

Incentives: Push vs. Pull

Incentives used to engage the participation of commercial parties are generally thought of as either "push" or "pull" incentives, with push funding inputs, and pull funding or rewarding outputs. Push strategies should focus on cultivating partnerships and collaborations, James Guyton of PRTM said, while pull strategies should focus on increasing market sustainability. Guyton highlighted a number of push and pull incentives. Common incentives for industry are listed in Box 9.

A recent study by the London School of Economics supported the use of a *combination* of push and pull incentives (LSE, 2010). The study, described by Chantal Morel of the London School of Economics and Political Science, looked at incentives to spur research and development for antibiotics, specifically those for which there are high levels of pathogen resistance, and considered 24 types of incentive mechanisms.

As discussed above, pharmaceutical companies will assume the risk of failure in development if success can be expected to bring substantial market return. The industry is accustomed to raising capital to fund research when they choose to assume that risk. In that regard, some participants commented that industry does not need push incentives, but rather an assurance of a market (e.g., via a pull incentive). A report by BIO Ventures for Global Health supports the use of market-based incentives, in particular, advanced market commitments (BVGH, 2006). A workshop participant cautioned that paying incentives to industry up front (push incentives) raises the issue of the government "paying for failures." Adopting a defense contractor model, where the government assumes all risk, is the most expensive and least productive way to proceed. Goal-line benefits (e.g., cash, patent benefits, liability protections) are a better approach.

Rex of AstraZeneca said, however, that push incentives are useful and appropriate for industry. Researchers, he said, must compete internally for company resources. If there is a tax incentive available for research on products for a particular threat, it is easier to make the business case internally that such research should be brought on board. Another participant noted that tax incentives are beneficial primarily for profitable companies, and therefore the majority of small companies that are innovating will not perceive tax credits as an incentive.

Offering a small-company perspective, Martin of Dynavax supported the concept of a prize-based funding strategy (retrospectively funding based on accomplishment of critical elements of a product profile), noting that such an approach creates a tremendous timing incentive. "If I can achieve the prize in one year by doing the critical experiment to show I have the appropriate attributes of the product profile, that is more interesting to me than achieving the same prize in 3 years," Martin said.

BOX 9
Incentives That Could Be Applied in the Development of Medical Countermeasures

Push incentives focus on cultivating partnerships between government and companies:

- Tax credits and grants
  - Helps address financing constraints; standard product risk still present.
- Public–private partnerships
- Access to intellectual property
- Minimal disincentives
- Regulatory clarity
- Contracting flexibility
- Sufficient and sustained funding
- Top-down leadership starting at the White House

Pull incentives focus on market sustainability:

- Advanced market commitments
  - Legally enforceable price commitments that are conditional on demand. Firms only see this price if there is demand. Addresses time-inconsistency risk. Leaves some demand risk on firms.
- Priority review vouchers
  - Of potential value to company planning to bring another product to market. Value of resale by a small company unable to use for subsequent product not yet established.
- Market exclusivity rewards and patent extensions
  - As medical countermeasures have very small, variable, or even nonexistent markets, exclusivity may not be an incentive.
- Prizes in exchange for intellectual property/license
  - Fee is conditional on development, regardless of demand.
  - Government absorbs demand risk, project risk by firms.
  - May provide a timing incentive.
- Patent extensions, data protection
- Liability protection

Regarding market exclusivity as an incentive, it was noted that medical countermeasures have very small, variable, or even nonexistent markets, and such exclusivity will not give rise to a strong private-sector reaction.

A question was also raised regarding how to encourage studies of appropriate countermeasure use in special populations, such as children. Although a pediatric incentive[11] exists to encourage the testing of

---

[11] Under the "Pediatric Rule," FDA can issue a written request to the sponsor of an investigational product, or the manufacturer of an approved product, that

products in children, in many situations involving countermeasures development, there will not be clinical trials at all, negating the current pediatric incentive.

Russell added that operationalizing many of these incentives is a real management challenge because they can be difficult to administer equitably and effectively.

Priority Review Voucher

The Priority Review Voucher was also discussed in some depth. As David Ridley of Duke University explained, this FDA program was enacted into law in September 2007 (Public Law 110-85) to establish an additional incentive for companies to invest in new drugs and vaccines. Under the program a company that develops a treatment for a neglected disease is awarded a Priority Review Voucher, which it can then (1) use when submitting an approval application for a different drug, or (2) sell to another company. Redeeming a voucher and receiving priority review would theoretically allow a company to bring a potential blockbuster drug to market sooner. Ridley noted that Novartis recently received the first voucher, but has not applied it yet. In support of the Priority Review Voucher, Smith of UPMC said there is a potential for significant benefit at minimal social cost.

A participant from a small company, however, expressed skepticism about the value of the Priority Review Voucher. As a small business enterprise, her company would be able to obtain a Priority Review Voucher, but most likely could not use it for the company for another product. Because no company has redeemed one yet, it is unclear what value it would hold for a large pharmaceutical company if the small company would try to sell it. Ridley acknowledged that until the vouchers begin to be used, there is great uncertainty associated with their value.

Incentives Not Needed?: Making a Strong Business Case

The government needs to try to speak the same language as industry when seeking collaboration. When approaching companies, the answer given is a function of the question being asked, said Douglas Pon, Pfizer's assistant vice president, Licensing in Global Business Development.

---

studies in children be conducted. The reward for conducting the requested studies is a 6-month period of marketing exclusivity ("pediatric exclusivity").

Industry is consolidating, cutting back, and trying to become more efficient. Companies are assessing their portfolios, looking at medical need, technical feasibility, and regulatory hurdles. When industry looks at medical countermeasures in terms of the larger company portfolio, it sees low margin and low volume. Consequently, the U.S. government needs to establish a very good business case, with a solid product profile including, for example, the target population, mechanisms of action, and forecasted demand, Pon commented. This business case should also provide a gap analysis, defining what technologies are needed, what other intellectual property will be needed, and what would be required in terms of scale-up process development and manufacturing.

All of this could be done through a government-backed "incubator" or center of excellence, similar to how venture capital investors foster innovation, Pon said. Venture capitalists identify a need and identify the technology and intellectual property that are required, and they build a sound and strong scientific and operational management team. For any gaps, they form strategic partnerships. This is how the government may be able to engage the pharmaceutical industry. The government needs to approach senior management with a direct question and a defined plan to be able to expect a direct and sincere answer in response.

Pon said incentives may not be necessary if there is a true and urgent national need for these products. Whittle the long list of potential threats down to a priority few, define what is actually needed in terms of manufacturing, analytic support, and other capabilities, and come to the table with proposals ready for discussion. Follow the same model that companies use internally, Pon said, by approaching corporate management with a sound business case and request for resources.

An RFP-style process, Pon said, where government publishes a notice that there is an opportunity for collaboration, is not going to work. *Government leadership* needs to engage individual *corporate leaders* directly, with a plan in hand, and ask for help.

NEW PARADIGMS, STRATEGIES, AND TACTICS FOR ENHANCING THE COUNTERMEASURES DEVELOPMENT ENTERPRISE

In addition to the models and strategies discussed above that have already been applied to medical countermeasures development (e.g., central oversight/single decision authority, TMTI, Incident Command System; Unmet Medical Need/Dual-Use approach, countermeasures as orphan drugs) and the models that have been successful in other venues (e.g., shared-risk strategies, open innovation, public–private partnerships, consortia, government as a strategic investor), participants discussed a host of other potential paradigms, strategies, and tactics for consideration.

Outsourcing Program Management

Several participants suggested variations of a paradigm where the government would outsource the facilitation responsibilities associated with the coordination of countermeasures development activities to a single entity. Venkayya of the Gates Foundation said there are many capacities that small and medium-sized companies do not necessarily have, and it does not make sense for the U.S. government to build the capacity internally to manage the many industry partners and their needs. The government needs to set requirements and expectations, and be responsible for portfolio management, but not project management.

Robert House, president, DynPort Vaccine Company LLC, offered the concept of an *integrator* function that would be responsible for program management, contract management, and risk management (Figure 1). To manage risk, the integrator must be aware of both the technical challenges as well as the programmatic and logistical challenges. The integrator needs to be "agnostic," that is, it must have no vested interest in seeing any particular technology taken forward. For any given need, the integrator would be responsible for pulling together all of the various components necessary: non-clinical, clinical, manufacturing, or other identified gap areas. The integrator would be staffed by technical experts and thus, for this approach to be successful, several participants said the government has to be willing to pay for a high-quality operation.

There can be regular engagement among the FDA, NIH, BARDA, and CDC, and this program coordination entity, as well as involvement of expert advisory boards and consultants. The government retains portfolio management responsibility, stewardship responsibilities, and, ultimately, go/no-go decisions.

Advantages of such an approach are that multiple candidates could be continuously fed into this system, subcontractors could be chosen based on their expertise, and the project therefore need not be tied to any one company's expertise. The system would allow technology innovators to concentrate on doing what they do best, which is discovery and early development of new technologies. Not all small companies want to be large companies, said House of DynPort Vaccine Company, and this approach would allow the small companies to continue to feed innovative product candidates into the countermeasures enterprise. This type of system would limit the need for BARDA to be deeply involved in managing the project and technology providers, so it could focus instead on overall portfolio success.

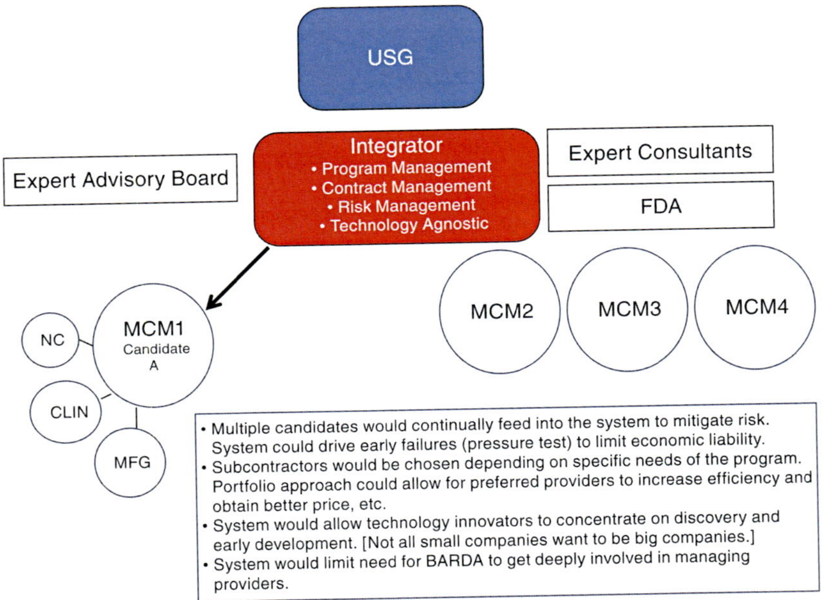

FIGURE 1 Model of medical countermeasures (MCM) development incorporating an integrator function. The U.S. government (USG) sets requirements and expectations and is responsible for portfolio management. The Integrator is responsible for project management.

NOTE: CLIN = clinical; MFG = manufacturing; NC = non-clinical.

SOURCE: Robert House. 2010. Presentation at IOM Workshop; The Public Health Emergency Medical Countermeasures Enterprise: Innovative strategies to enhance products from discovery through approval.

Government as a Strategic Partner

In addition to the "government as strategic investor" approach (à la In-Q-Tel) discussed above, Monath of Kleiner Perkins Caufield & Byers also suggested government approach industry as a *strategic partner*, looking to acquire technology, bring it in-house, and develop it. In this regard, government needs to provide small biotechnology companies with deals in terms that are familiar to them and the commercial space. For example, end-license the technology, provide upfront payments, and provide milestones when it reaches the appropriate point in its development. If there is procurement, and if doses are put into the Strategic National Stockpile, there is a royalty or an equivalent at the end. Monath said that biotechnology companies would be interested in seeing their technology used for that purpose by somebody else, in return for a deal in commercial terms.

Platform Technologies

There was much discussion of the development and use of platform technologies, as well as FDA approval of products developed on a new or established platform. Technology platforms can help address the unknown need, those threats and emerging pathogens that cannot be anticipated, meaning we cannot vaccinate against them.

The countermeasures enterprise should have a focus on multiuse products and multiuse platform technologies that span the entire public health product-development portfolio (i.e., existing and emerging public health problems and deliberately released public health problems), NIAID's Fauci said. Focus should be on development of new platforms as opposed to going after a particular threat. As an example, Fauci cited a recent publication describing a broad-spectrum antiviral drug targeting enveloped viruses, noting that this could be effective against yet-to-be discovered enveloped viruses (Wolf et al., 2010).

Although the idea of platform technologies is attractive, the regulatory environment needs to be prepared to swiftly approve new platform technologies and to approve products developed on those platforms. As discussed previously, a clear regulatory pathway must be established for medical countermeasures. The first product through on any given platform is indeed a new product. The question then is to what degree second and subsequent products using the same platform can extrapolate data (perhaps similar to how the abbreviated new drug applications used for generic drugs can include data extrapolated from the pioneer product), commented Cassell, chair of the workshop.

One caution noted by Gomez of PRTM is that while it is fairly easy to know when a product has failed, it can be very difficult to tell when a platform has failed because there is always an improvement, always a new disease target.

Revised PHEMCE Implementation Plan

As discussed above, a challenge for industry is that the PHEMCE implementation plan itself provides limited guidance in that it is essentially a list of pathogens and agents of interest to the government. Rose proposed a use-based medical countermeasure acquisition matrix (Figure 2). The concept of operations is similar across all of the organisms listed. There is a need for pre-event prophylaxis, particularly for the military and first responders. Treatments are needed for people who develop symptomatic illness when an outbreak occurs. For those who are exposed but not yet ill, postexposure prophylaxis is needed to keep them from developing illness. Rose noted that the matrix is for healthy adults and a similar matrix for children or the elderly would show significantly more yellow and red.

Threat	Pre-Event Prophylaxis	Treatment	Post-exp. Prophylaxis	General Prophylaxis
Anthrax	Vaccine	Antibiotics, MoAb	Antibiotics +/– Vaccine	Vaccine
Smallpox	Vaccine	None	Vaccine	Vaccine
Plague	None	Antibiotics	Antibiotics	None
Tularemia	None	Antibiotics	Antibiotics	None
Viral Hem. Fevers	None	None	None	None

FIGURE 2 Use-based medical countermeasure acquisition matrix (Healthy Adults 18–64). Green: product is available; yellow: product is not yet approved; red: no products for development.

SOURCE: Eric Rose. 2010. Presentation at IOM Workshop; The Public Health Emergency Medical Countermeasures Enterprise: Innovative strategies to enhance products from discovery through approval.

EXISTING REGULATORY TOOLS AND APPROACHES THAT CAN BE APPLIED TO ADVANCE COUNTERMEASURES DEVELOPMENT

Boris Lushniak, assistant commissioner for Counterterrorism Policy at FDA, said the agency's mission is very clear: safety, efficacy, and quality. Lushniak and Jeanne Novak, of CBR International Corp., both drew attention to policies and programs already in place that could be used to facilitate medical countermeasure development. However, throughout the workshop, it was repeatedly noted that there is a need for a clear regulatory pathway necessary to evaluate and regulate the biodefense products as well as a number of additional gaps and associated opportunities.

Opportunities for Accelerating Approval of Medical Countermeasures: Evolving the Regulatory Framework

Numerous individual suggestions were made about addressing the regulatory aspect of the medical countermeasures enterprise. They are compiled here as part of the factual summary of the workshop, and should not be construed as reflecting consensus or endorsement by the workshop, the Forums, or The National Academies. They are as follows:

- **Fund, support, and enable regulatory science.** Develop, assess, and provide tools, methods, models, standards, guidance, and pathways to evaluate product safety, efficacy, and quality (e.g., biomarkers; surrogate endpoints; adaptive and other flexible clinical trial designs; rapid scale-up of production; rapid methods to assess purity, potency, quality, contamination).

- **Declare medical countermeasure to be a priority at the FDA,** which would entail providing security clearances to senior FDA employees and briefing all medical officers across the entire agency about the real risks and the immediate need for products.

- **Create a designation indicating that a program is relevant to national security and establish priority review for medical countermeasure applications.** For products so designated, FDA can then provide additional assistance, perhaps some form of priority review or acceleration of time lines consistent with the product being part of a national security countermeasures development program.

- **Consider creating an alternative approval mechanism for medical countermeasures** with criteria more stringent than EUA and less stringent than normal licensure for commercial products.

- **Formalize the EUA process.** The existing EUA mechanism works well as an emergency response process, but it is not a preparedness mechanism. A formal EUA approval process would use existing criteria for the EUA, with the exception of the actual declaration of an emergency, and would differ from the current process, in which submission of pre-EUA data is encouraged, but no response is required, and typically no response is generated.

- **Balance data needs according to risk/benefit.** Because medical countermeasures will be used in situations of grave risk, the amount of data needed for approval may be less than that for a more mainstream product. Early HIV interventions were cited as a parallel situation.

- **Facilitate communication with product sponsors.** In-person meetings and site visits between the FDA and with companies developing medical countermeasures could be done with increased frequency and flexibility, without having to formally request the existing FDA A, B, and C meeting categories.[12]

- **Send FDA agency staff into the field to assist companies as needed.** This was done, for example, when Genentech was developing the first recombinant human growth hormone, a product for which there was urgent need. An FDA senior chemist spent 3 months at Genentech, working with its staff to prepare manufacturing for approval.

- **Abandon the Draft Guidance on the Animal Rule,** which is significantly more restrictive than the Animal Rule itself. The Rule says FDA may grant approval when "the results of those animal

---

[12] Through the 1997 FDA Modernization Act, the FDA established three main categories of meetings—A, B, or C—that describe the various types of formal industry meetings that occur (http://www.fda.gov/Drugs/DevelopmentApprovalProcess/FormsSubmission Requirents/ElectronicSubmissions/DataStandardsManualmonographs/ucm071774.htm).

studies establish that the drug is reasonably likely to produce clinical benefit in humans."

- **Emphasize emergency preparedness throughout the FDA.** Instill cultural change though an agency-wide focus on emergency preparedness/national security in the regulatory environment, and encourage use of many of the tools that already exist in the current regulatory laws and regulations, in a more efficient, focused way.

- **Provide regulatory and licensure guidance to funding agencies on development programs.** FDA personnel can assist funding agency (HHS and DoD) reviews of medical countermeasures development programs. This would facilitate progress by providing a "unified" development method. Specialized Review and Development Teams could be formed at the FDA (center and division levels). Funding agencies would have access to the Specialized Review and Development Teams in order to seek expert FDA opinion about progress of a particular development program.

- **Reflect FDA data needs in NIH trial design and conduct.** NIH funds many large clinical trials, which are then incorporated into an application for product approval. Many times the data are not adequate to support FDA decision making in terms of safety, efficacy, and labeling claims. If FDA is at the table with NIH when these trials are being planned, and when they are being executed, they can make sure the supporting data are available when needed for regulatory decisions.

The Way Forward: Themes from the Workshop

Additional suggestions regarding a wide variety of areas relevant to medical countermeasures development were presented by individuals during the workshop and are compiled here. Investigating details about the feasibility and implementation of these ideas were beyond the scope of the workshop. The additional suggestions are the following:

- **Institute a single, unified management structure for the federal countermeasures enterprise.** The hallmark of a successful developmental program is strong leadership that can have a singular focus on developing products—a system that could direct basic and applied research and have the power to force alignment of all the components. BARDA currently does not have this authority. However, consideration should be made for the complementary, but separate, missions of HHS and the DoD.

- **Facilitate improved end-to-end partnering** between federal agencies (NIH, ASPR/BARDA, DoD, FDA), industry, and academia throughout research and development of medical countermeasures.

- **Declassify threat analysis data to allow transparency for potential product requirements,** and thus to improve the understanding among the public, legislators, and industry about the scope of the threat.

- **Provide additional venues for classified briefings for all segments of the countermeasures enterprise, including private-sector stakeholders.** Two examples are the Joint Civilian Orientation Conference sponsored by DoD and the FBI's National Security Higher Education Advisory Board, where leaders in business and academia get security clearance, are briefed on issues, and contribute to discussions on national security matters. Something comparable could be established for the leadership from pharmaceutical and biotechnology companies.

- **Develop target product profiles.** Define the product, including both the indication (i.e., disease/condition to be treated) and what an appropriate product would be to use in a public health emergency. Prospectively establish the metrics that define success for a product, then build an experimental plan that assesses whether a product has critical attributes.

- **Utilize grants and contracts to incentivize research and development in multi-use agents.** BARDA, and other government grants and contracts, as well as milestone payments serve as important incentives for the development of multi-use agents, in particular for agents like antibiotics, which do not always have a very high profit margin.

- **Empower program managers.** Provide increased authority to program managers to develop a vision of what a successful project or program encompasses, and then to go out and secure the resources necessary to put that vision together. Small companies would benefit from project managers and/or FDA being able to reach out proactively at the beginning of a contract to discuss how the company could avoid potential hurdles and common mistakes related to complex areas such as the Animal Rule, GLP toxicology studies, etc.

- **Use the Federal Acquisition Regulation to create commercial markets for biodefense products.** FAR § 12.102(f)(1) states that contracting officers "may treat any acquisition of supplies or services that, as determined by the head of the agency, are to be used to facilitate defense against or recovery from nuclear, biological, chemical, or radiological attack, as an acquisition of commercial items." This establishes the authority to designate a countermeasure as a commercial item (not a cost-plus item) that can be compared to other drugs and vaccines already in the commercial markets.

- **Simplify the intellectual property system and improve alignment between FDA approval and when a patent is granted.** Ascribe value to the data themselves. The data are what determine whether the product is of any value to the public. Consider a full 20-year patent term for government-chosen medical countermeasures and a data protection period of 20 years, both running from the day of FDA approval.

- **Establish incentives and public private partnerships to encourage greater investment by the private sector.** Particular focus should also be made on how to incentive the development of countermeasures for children and other special populations.

- **Ensure integrity of Project Bioshield funds and increase funding for countermeasures research and development.** It is important to guarantee that procurement resources remain available, but far greater resources are currently required to ensure necessary countermeasures research and development. Removing funds from Project Bioshield's Reserve Fund for research and development could have serious repercussions on the government's ability to guarantee a suitable marketplace for future countermeasure procurement. It also would contribute to uncertainty among the private sector about long-term, stable funding for countermeasures research and development.

- **Commit to multiyear federal funding of CBRN/pandemic countermeasures to ensure companies are able to maintain a continuous development program.**

- **Consider a tax benefit for equipment purchased for use in the development or manufacturing of medical counter-measures.** The Department of the Treasury could approve an accelerated depreciation schedule for equipment that is required for biodefense products exclusively (i.e., that has to be separate from the rest of the equipment in a research and development or manufacturing facility).

- **Make the federal government the defendant in all the tort claims resulting from countermeasures.** The government is the defendant in the Vaccine Injury Compensation Act, but not in the existing PREP Act. Legislatively eliminate the cause of action and replace it with this administrative compensation system.

- **Make suspension of state liability in a federally declared emergency a condition for receipt of biodefense money for the states.** The federal government cannot force states to pass legislation, but it can withhold preparedness funding if the state fails to comply with the conditions of the grant.

- **Institutionalize capacities.** Planning is important, but adaptability and rapid responses are critical elements for success. Integrate problem solving and flexible approaches into countermeasure development and emergency planning. Institutionalize capacities throughout the countermeasures enterprise rather than reserve them for a crisis. The Illinois Department of Public Health, for example, sent 42 senior staff members to be trained in the National Incident Management System, providing them with useful background on an organizational approach to managing problems. These skills were of great value during the 2009 H1N1 pandemic.

- **Sponsor a contracting office training course,** covering best practices to address countermeasure development, licensure, manufacturing, and procurement with biodefense, and updating officers on available options.

- **Establish more agile platforms and multiuse technologies** (e.g., vaccine, diagnostic, or monoclonal platforms) that can be rapidly adaptable to address new pathogens.

- **Engage end users, specifically public health professionals and healthcare providers, in requirement setting.** Incorporating end-user perspectives in the requirement setting for product development will help ensure that the final product that is stockpiled or distributed will be able to be administered effectively and efficiently to the target population. In this regard, the needs of special populations, including children, are part of planning, research, and development.

- **Improve public engagement.** Robust and continued public engagement helps to ensure the legitimacy of the counter-measures enterprise, communicates the risk, communicates those plans currently in place, and helps produce the best possible result. Engage faith-based institutions. Use the educational system; develop a message that becomes part of a school curriculum. Work to dispel myths and misperceptions.

- **Create technical centers of excellence** to help ensure the necessary expertise, including dedicated capabilities and core resources and manufacturing facilities.

- **Invest in career development strategies to ensure the necessary scientific and regulatory expertise.** Define a curriculum at universities that could support the countermeasures enterprise. The government could sponsor educational opportunities to improve human capital capacity. For example, a fellowship for graduate degree candidates established by FDA and BARDA could enrich the countermeasures enterprise and bring new people in the system.

CONCLUSION

Deliberate acts of bioterrorism and emerging natural infections will continue to be a major public health concern. There is no one specific right incentive, or one specific model that is best suited to advanced development of countermeasures against chemical, biological, radiological, and nuclear agents. No single player is ideally positioned to meet the needs of the medical countermeasures enterprise.

There are at least two different ideological approaches to the development of medical countermeasures, as follows:

- Consider it to be a critical national security issue, with an urgent need to develop countermeasures for an imminent threat.
- Focus on addressing important unmet medical needs that are present today, and while these needs are somewhat different, some benefit can be derived for the national security mission in the end.

These approaches are not mutually exclusive, and the ideal countermeasures enterprise would have a multiuse focus on existing, emerging, and deliberately released threats. However even if cross-indications are feasible, a national threat may, or likely will, occur without notice. Therefore, without emphasis upfront on the national security implications at the federal level on development pathways, medical countermeasures are less likely to be available when needed. The use of these two approaches in combination could provide the basis of a robust development strategy and would allow federal systems to be in place to deal with imminent threats, in distinction from standard drug development processes.

Assistant Secretary Lurie reminded attendees that engaging industry in the countermeasures enterprise is essential. A dramatic transformation is needed in the way the federal government interacts with industry. Government and industry must come together as trusted partners, and this relationship must be sustained—not only in a time of a crisis. There is no one push or pull incentive that is sufficient to attract experienced companies to the enterprise. Similarly, there is no "silver bullet" combination of incentives that suits all situations.

Seeking private-sector participation in developing these products is not about saving the public money. It is about bringing innovation, human capital, entrepreneurship, and private-sector expertise to bear on important public health issues. Market incentives cost money. The U.S. government should be willing to pay for countermeasures, to the extent that there are national security and public health needs.

Throughout the workshop it was frequently asserted that partners in the countermeasures development enterprise should be able to expect the following:

- A clear articulation of the threats
- A very clear articulation of the target product profile (indication, formulation, manufacturing needs, predicted demand)
- Clear time lines (what activities are on an urgency timeline vs. a long-term timeline)
- Predictability of finance (e.g., if a company delivers according to specifications, there will be a purchaser on the other end, and that purchaser will be the U.S. government)
- Predictability from the regulatory environment

Regardless of the approach taken, all participants in the medical countermeasures enterprise should remember the end goal: ensuring the health of the people. The medical countermeasures research and development enterprise should continue to develop and implement models that, while respecting industry and governmental needs and goals, foster collaborative relationships among government and industry to ensure maximum use of their respective strengths and capabilities for the protection of the nation's health.

A

References

BVGH (BIO Ventures for Global Health). 2006. *Advance market commitments to stimulate industry investment in global health product development.* http://bvgh.org/LinkClick.aspx?fileticket=W_5VAgleR8k%3d&tabid=91 (accessed March 18, 2010).

Center for Arms Control and Non-Proliferation: Scientists working group on biological and chemical weapons. 2010. *Biological Threats: A Matter of Balance.* Statement released January 26, 2010. http://www.armscontrolcenter.org/policy/biochem/articles/biological_threats_a_matter_of_balance (accessed March 8, 2010).

FDA (Food and Drug Administration). 2004. *Innovation or stagnation: Challenge and opportunity on the critical path to new medical products.* http://www.fda.gov/downloads/ScienceResearch/SpecialTopics/CriticalPathIniative/CriticalPathOpportunitiesReports/UCM113411.pdf (accessed March 4, 2010).

FDA. 2007. *FDA science and mission at risk.* Report of the Subcommittee on Science and Technology. http://www.fda.gov/ohrms/dockets/AC/07/briefing/2007-4329b_02_01_FDA%20Report%20on%20Science%20and%20Technology.pdf (accessed March 9, 2010).

Graham, B., and J. Talent. 2009. *Bioterrorism: Redefining prevention. Biosecurity and Bioterrorism: Biodefense strategy, practice, and science.* 7(2):125–126. http://www.liebertonline.com/doi/pdfplus/10.1089/bsp.2009.0610?cookieSet=1 (accessed March 15, 2010).

IOM (Institute of Medicine). 2008. *The U.S. commitment to global health: Recommendations for the new Administration.* Washington, DC: The National Academies Press.

IOM. 2009. *Venture philanthropy strategies to support translational research: Workshop summary.* Washington, DC: The National Academies Press.

IOM and NRC (National Research Council). 2004. *Giving full measure to countermeasures: Addressing problems in the DoD program to develop medical countermeasures against biological warfare agents.* Washington, DC: The National Academies Press.

LSE (The London School of Economics and Political Science). 2010. *Policies and incentives for promoting innovation in antibiotic research.* http://www2.lse.ac.uk/newsAndMedia/news/archives/2009/09/antibiotics.aspx (accessed March 14, 2010).

Matheny, J., M. Mair, and B. Smith. 2008. Cost/success projections for U.S. biodefense counter-measure development. *Nature Biotechnology* 26(9):981–983.

Melese, T., S. M. Lin, J. L. Chang, and N. H. Cohen. 2009. Open innovation networks between academia and industry: An imperative for breakthrough therapies. *Nature Medicine* 15:502–507.

NBSB (National Biodefense Science Board). 2010a. *Where are the Countermeasures? Protecting America's Health From CBRN Threats.* http://www.hhs.gov/aspr/barda/documents/phemceimplplan_041607final.pdf (accessed May 17, 2010).

NBSB. 2010b. *Optimizing industrial involvement in medical countermeasure development.* http://www.hhs.gov/aspr/omsph/nbsb/nbsbrpt-2010.pdf (accessed March 2, 2010).

PhRMA. 2010. *R&D investment by U.S. biopharmaceutical companies remains strong despite ongoing economic challenges.* http://www.phrma.org/news/news/rd_investment_us_biopharmaceutical_companies_remains_strong_despite_ongoing_economic_chall (accessed March 30, 2010)

Wolf, M. C., A. N. Freiberg, T. Zhang, A. Akyol-Ataman, A. Grock, et al. 2010. A broad-spectrum antiviral targeting entry of enveloped viruses. *Proceedings of the National Academy of Sciences* 107(7):3157–3162.

B

Workshop Agenda

Day 1
February 22, 2010
House of Sweden
Anna Lindh Hall
2900 K Street, NW
Washington, DC 20007

Background:
This workshop will examine federal policies and activities that relate to discovery through approval of medical countermeasures (e.g., vaccines, drugs, and diagnostics) for responding to public health emergencies with the potential to rapidly overwhelm the public health and medical systems (e.g., terrorism and pandemic flu). The workshop will explore potential opportunities to enhance this enterprise by evaluating other models or systems that have similar goals of developing medical products with low commercial viability. As charged by the Secretary of Health and Human Services, the Office of the Assistant Secretary for Preparedness and Response is leading a review of the entire public health emergency medical countermeasures enterprise, to be completed in the first quarter of this year. Because there is a limited commercial market for most medical countermeasures, the government has had to create incentives to encourage private-sector pharmaceutical and biotech companies to develop the needed products. The countermeasures research and development (R&D) enterprise encompasses many partners from across the federal government, states, and industry, which need to function together to develop the medical countermeasures necessary to sustain

national health security. However, certain structural, strategic, and technical elements of the enterprise continue to impede research, development, and production of medical countermeasures. The presentations and discussions at this workshop are intended to assist federal officials in conducting a thorough review of the "pipeline through approval" spectrum of our national programs and to assist in the ultimate goal of improving the efficiency and effectiveness of the countermeasures enterprise.

Meeting Objectives:

- Identify and discuss strategies to optimize the federal public health emergency medical countermeasures enterprise, and explore resources and/or other supporting components needed for accomplishing goals of countermeasure discovery, development, approval, and production.
- Examine strategies to further enhance the translation of early phase investments in basic science into potential public health interventions.
- Identify and discuss models for enhancing current partnerships and establishing new ones among federal programs, innovators, and the commercial marketplace to enhance our nation's capabilities to meet public health emergency preparedness goals.
- Consider market forces acting on the advanced development biodefense community (pharma/biotech) that incentivize/disincentivize efforts to develop and license products in support of the national response.
- Examine ways the regulatory oversight process for public health emergency medical countermeasures might evolve and identify ways to enable more efficient approval and use.
- Review the innovative approaches being used to advance drug development for orphan diseases (i.e., rare, neglected, or tropical diseases) or any other area that does not have a ready and sustainable commercial market (e.g., oncology therapeutics), and identify the shared challenges and opportunities for strategies that might be adopted by the Enterprise.

Working Dinner

6:00 p.m. Welcome, Introductions, and Charge to Workshop Participants

GAIL CASSELL, *Workshop Chair*
Vice President, Scientific Affairs and Distinguished Lilly Research Scholar for Infectious Diseases
Eli Lilly and Company

6:15 p.m. Needs and Opportunities to Advance the Countermeasures Enterprise

RANDY LARSEN
Executive Director, Commission on the Prevention of Weapons of Mass Destruction Proliferation and Terrorism

Presentation of White Papers

6:35 p.m. Case Studies of the HHS Medical Countermeasure Programs (Paper 1)

ROBERT KADLEC
Vice President
Global Public Sector
PRTM

6:55 p.m. Optimizing the Medical Countermeasure Product Pipeline from the Science Base Through Advanced Development (Paper 2)

GEORGE KORCH
Senior Science Advisor
Principal Deputy Assistant Secretary for Preparedness and Response
Office of the Assistant Secretary for Preparedness and Response
Department of Health and Human Services

7:15 p.m. Synthesis of Business Models and Economic and Market Incentives for Vaccines and Therapeutics (Paper 3)

JAMES GUYTON
Principal
Public Health and Biodefense Practice
PRTM

7:35 p.m. National Biodefense Science Board (NBSB): Medical Countermeasure Markets and Sustainability

JOHN GRABENSTEIN
Senior Medical Director for Merck Vaccines
Member, National Biodefense Science Board

7:50 p.m. Discussion with Attendees

8:15 p.m. ADJOURN

Day 2
February 23, 2010
House of Sweden
Alfred Nobel Hall
2900 K Street, NW
Washington, DC 20007

8:00 a.m. Welcome and Introductions

GAIL CASSELL, *Workshop Chair*
Vice President, Scientific Affairs and
Distinguished Lilly Research Scholar for
Infectious Diseases
Eli Lilly and Company

8:10 a.m. Charge to Workshop Participants and Overview of the Federal Public Health Countermeasures Enterprise: Challenges and Opportunities

NICOLE LURIE
Assistant Secretary for Preparedness and Response
Office of the Assistant Secretary for Preparedness and Response
Department of Health and Human Services

8:30 a.m. The Public Health Perspective on Medical Countermeasure Development, Acquisition, and Use

THOMAS FRIEDEN
Director
Centers for Disease Control and Prevention

SESSION I: THE COUNTERMEASURES ENTERPRISE: OVERVIEW OF THE CHALLENGES AND OPPORTUNITIES

8:50 a.m. Past and Present Enterprise Efforts: Why Are We Where We Are and What Models Are Most Likely to Succeed?

PHILIP RUSSELL
Board of Trustees
Sabin Vaccine Institute

9:05 a.m. International Approaches to Countermeasure Research, Development, and Approval

MARIA JULIA MARINISSEN
Team Leader, International Partnerships and Initiatives
Office of the Assistant Secretary for Preparedness and Response
Department of Health and Human Services

9:20 a.m. Defining the Steps of the Critical Pathway: Strategies to Move Forward

MARIETTA ANTHONY
Associate Director
Arizona Center for Education & Research on Therapeutics
Critical Path Institute

9:35 a.m. Commercial Challenges: Perspectives from the Biotech Industry

ERIC ROSE
CEO and Chair, Board of Directors
Siga Technologies, Inc.

9:50 a.m. Commercial Challenges: Perspectives from Big Pharma

JOHN REX
Infection Clinical Vice President
Oncology & Infection Therapy Area
AstraZeneca

10:05 a.m. Discussion with Attendees

10:30 a.m. BREAK

SESSION II: MEDICAL COUNTERMEASURE EXPERIENCES FROM PANDEMIC INFLUENZA AND ANTHRAX PLANNING: IDENTIFYING LATE-STAGE ENTERPRISE ISSUES THAT IMPACT EARLY STAGE DECISION MAKING

Session Objectives:

- Provide context for the countermeasures enterprise given the ultimate use of medical countermeasures.
- Explore how the considerations of various end users should inform the design of the Enterprise and developmental product profiles.

- Provide CDC and state public health practice experiences and lessons regarding H1N1 pandemic medical countermeasure policies, distribution, and uses.
- Provide examples of challenges and opportunities faced in developing an anthrax medical countermeasures distribution model.
- Identify the issues from H1N1 experiences that can be generalized or transferred to questions about the broad range of medical countermeasure programs.

10:45 a.m. Session Introduction

MONIQUE MANSOURA, *Session Chair*
Director for Medical Countermeasure Policy, Planning and Requirements
Office of the Biomedical Advanced Research and Development Authority, Office of the Assistant Secretary for Preparedness and Response
Department of Health and Human Services

10:50 a.m. Panel Presentations

DANIEL JERNIGAN
Deputy Influenza Director
Centers for Disease Control and Prevention

SUSAN COOPER
Commissioner
Tennessee Department of Public Health

DAMON ARNOLD *(via telecon)*
Director
Illinois Department of Public Health

ANDREW PAVIA
George and Esther Gross Presidential Professor
Division of Pediatric Infectious Diseases
University of Utah School of Medicine

11:10 a.m. Discussion with Attendees

11:40 a.m. LUNCH: Atrium Lounge

SESSION III: OPTIMIZING THE RESEARCH AND DEVELOPMENT ELEMENTS OF THE ENTERPRISE

Session Objectives:

- Review the structural and strategic elements of current efforts by the relevant federal agencies to support the research, development, and production of emergency medical countermeasures, including consideration of the underlying infrastructure that supports these efforts.
- Identify and discuss the critical paths and systems approaches needed to optimize the research, development, and approval elements of the medical countermeasures development enterprise.
- Propose strategies to optimize the federal public health countermeasures enterprise.
- Identify and discuss models for enhancing collaboration and coordination among relevant federal programs. Examine the enabling technologies and infrastructures that will be necessary.

12:30 p.m. Session Introduction and Objectives

ROBERT KADLEC, *Session Chair*
Vice President
Global Public Sector
PRTM

12:35 p.m. Panel Discussion: The Countermeasures Enterprise: Current Constraints and Opportunities to Move Forward

- Review the structural and strategic elements of current efforts by the relevant federal agencies to support the research, development, and production of emergency medical countermeasures, including consideration of the underlying infrastructure that supports these efforts.
- Discuss the current constraints of the structure and organization of the countermeasure enterprise.

- Explore opportunities to optimize the public health emergency medical countermeasures enterprise.
  - Propose strategies for enhancing collaboration and coordination within the enterprise, taking into account constraints related to ongoing programmatic activities, budget cycles, overlapping goals, and limitations due to existing federal regulations and departmental policies.

NIH

ANTHONY FAUCI
Director
National Institute of Allergy and Infectious Diseases, NIH
Department of Health and Human Services

BARDA

ROBIN ROBINSON
Director
Office of the Biomedical Advanced Research and Development Authority, Office of the Assistant Secretary for Preparedness and Response
Department of Health and Human Services

DoD

KENNETH COLE
Medical Director
Nuclear and Chemical and Biological Defense Programs
Office of the Deputy Assistant to the Secretary of Defense
Department of Defense

Perspective of Prior Leadership

D. A. HENDERSON
Former Director, Office of Public Health Emergency Preparedness
Director of WHO Smallpox Eradication Program
Distinguished Scholar, Center for Biosecurity of UPMC

1:30 p.m. Discussion with Attendees

2:00 p.m. BREAK

SESSION IV: OPPORTUNITIES FOR ENHANCING TRANSLATION OF BASIC SCIENCE: MODELS TO IMPROVE INNOVATION THAT RESPONDS TO NATIONAL PRIORITIES

Session Objectives:

- Explore how our investment in basic research is currently "procured" and "exploited" to yield products for the advanced development pipeline.
- Identify potential other models to improve innovation based on national priorities.
- Examine whether the structure and nature of current investments in science lead to the intended result of potential products or product candidates.
- Discuss how the Enterprise can be the most collaborative and constructive partners with industry to ensure efficient product development.
- Discuss how the current investment in basic science can be better exploited to improve the MCM research infrastructure to ensure more coordinated and collaborative research that effectively advances the Enterprise.

2:15 p.m. Panel Discussion: Models for Procuring Science

MICHAEL KURILLA, *Panel Chair*
Director, Office of BioDefense Research Affairs
Associate Director for BioDefense Product Development
NIAID

MICHAEL GOLDBLATT
President and CEO
Functional Genetics

DAVID WURTMAN
VP, Corporate Development
NexBio

BRETT GIROIR
Vice Chancellor for Research
Texas A&M

PAUL OWENS
Senior Director, Chorus CMC
Lilly Research Laboratories

2:50 p.m. Discussion with Attendees

3:20 p.m. Panel Discussion: Strategies to Improve Portfolio Management: Translating Basic Science

PHYLLIS ARTHUR, *Panel Chair*
Director
Health & Regulatory Affairs
BIO

Overview: Current Process for Making Go/No-Go Decisions

MICHAEL KURILLA
Director, Office of BioDefense Research Affairs
Associate Director for BioDefense Product Development
NIAID

Panel Discussion

DAVID PERRYMAN
President & CEO
Zirus

GEORGE PAINTER
CEO
Chimerix

TYLER MARTIN
Chief Medical Officer
Dynavax

PATRICK IVERSEN
Senior Vice President
Strategic Alliances
AVI BioPharma

JOHN REX
Infection Clinical Vice President
Oncology & Infection Therapy Area
AstraZeneca

4:00 p.m. Discussion with Attendees

5:00 p.m. ADJOURN

Day 3
February 24, 2010
House of Sweden
Alfred Nobel Hall
2900 K Street, NW
Washington, DC 20007

KEYNOTE PRESENTATION: FDA CONTRIBUTIONS TO THE ENTERPRISE: CURRENT AND FUTURE STRATEGIES

8:00 a.m.	Welcome and Introduction GAIL CASSELL, *Workshop Chair* Vice President, Scientific Affairs and Distinguished Lilly Research Scholar for Infectious Diseases Eli Lilly and Company
8:05 a.m.	Keynote Presentation JESSE L. GOODMAN Chief Scientist and Deputy Commissioner Science and Public Health (Acting) Food and Drug Administration

SESSION V: OPPORTUNITIES FOR ACCELERATING APPROVAL OF MEDICAL COUNTERMEASURES: EVOLVING THE REGULATORY FRAMEWORK

Session Objectives: Review current regulatory authority and discuss current barriers to approval of products emerging from the countermeasures enterprise. Identify and discuss innovative approaches to facilitate effective regulation of countermeasures for rapidly emerging and/or rare public health threats, while still ensuring appropriate review of safety and efficacy data. Examine the current scientific infrastructure at the FDA and opportunities for the agency to be better prepared for the needs of the countermeasure community. Discuss whether FDA approval should be the standard for all medical countermeasures.

8:35 a.m. Session Objectives and Introduction

BORIS LUSHNIAK, *Session Chair*
Assistant Commissioner for Counterterrorism Policy
Food and Drug Administration

8:40 a.m. Panel Discussion

GERALD PARKER
Principal Deputy Assistant Secretary
Office of the Assistant Secretary for Preparedness and Response
Department of Health and Human Services

MARY PENDERGAST
Founder, Pendergast Consulting
Former Deputy Commissioner of the FDA

JEANNE NOVAK
CEO and President
CBR International Corp.

LUCIANA BORIO
Medical Reviewer
Office of Vaccine Research and Review
Center for Biologics Evaluation and Research
Food and Drug Administration

GAIL CASSELL
Vice President, Scientific Affairs and Distinguished Lilly Research Scholar for Infectious Diseases
Eli Lilly and Company

9:40 a.m. Discussion with Attendees

10:10 a.m. BREAK

SESSION VI: MARKET INCENTIVES IN THE DEVELOPMENT OF MEDICAL COUNTERMEASURES: IDENTIFYING MARKET OPPORTUNITIES AND ELIMINATING DISINCENTIVES

Session Objectives: How can the federal government better ensure that the emerging basic science concepts are translated into candidate countermeasures? What market issues exist, and what financial incentives are needed to overcome these issues? Explore the impact of the animal efficacy rule and opportunities to decrease its impact on the product development time frame while still ensuring appropriate review of safety and efficacy data. Discuss the advantages and disadvantages of establishing a consolidated federal intramural program to support translational and preclinical studies. Consider the impact of contract requirements and use policy on the private sector. Explore strategies to offset the limited commercial marketplace for emergency medical countermeasures. Examine the impact of existing liability provisions.

10:25 a.m. Session Objectives and Introduction

JOHN GRABENSTEIN, *Session Chair*
Senior Medical Director for Adult Vaccines
Merck & Co., Inc.

10:30 a.m. Panel Discussion: Other Models, Lessons Learned, and Success Stories from the World of "Orphan Drugs" and Challenging Commercial Products

CHANTAL MOREL *(via telecon)*
Department of Social Policy
London School of Economics

MARGARET ANDERSON
Executive Director
FasterCures

MELINDA MOREE
President and CEO
BIO Ventures for Global Health

MARLENE HAFFNER *(via telecon)*
President
Haffner Associates, LLC

VICTORIA SUTTON
Robert H. Bean Professor of Law
Texas Tech University School of Law

RAJEEV VENKAYYA
Director, Global Health Delivery
Bill & Melinda Gates Foundation

11:00 a.m. Discussion with Attendees

11:30 a.m. Panel Discussion: Market Incentives

WESLEY YIN
Assistant Professor
Department of Economics
Boston University

THOMAS MONATH
Partner
Pandemic and Biodefense Fund
Kleiner Perkins Caufield & Byers

DAVID GILBERT
Director
Infectious Diseases
Providence Health & Services

DAVID RIDLEY
Assistant Professor
The Fuqua School of Business
Duke University

BRUCE ARTIM
Director
Federal Affairs
Eli Lilly and Company

BRADLEY SMITH
Senior Associate
Center for Biosecurity, UPMC

12:15 p.m. Discussion with Attendees

12:45 a.m. LUNCH: Atrium Lounge

SESSION VII: THE ROLE OF PARTNERSHIPS AND ALTERNATIVE BUSINESS MODELS

Session Objectives: Examine how partnerships (public–private, private–private, and public–public) and alternative business models can be established to mitigate the risk for each sector. Discuss opportunities to leverage the expertise of pharma while retaining the innovation from biotech. Explore the role that public–private and private–private partnerships may have in supporting the development of multiproduct technologies and facilities. Identify opportunities for partnerships to help establish smoother transition from discovery to development and testing to regulatory approval for MCMs.

1:30 p.m. Session Objectives and Introduction

RONALD SALDARINI, *Session Chair*
President
Biological Initiatives

1:35 p.m. Panel Discussion: Partnerships: Opportunities to Leverage the Expertise of Pharma While Retaining the Innovation from Biotech

PHILLIP GOMEZ
Director
Global Public Sector
PRTM

TERI MELESE
Director of Research Technologies and Alliances
UCSF School of Medicine

ROBERT HOUSE
President
DynPort Vaccine Company, LLC

DOUGLAS PON
Assistant Vice President
Licensing in Global Business Development
Pfizer Inc.

2:05 p.m. Discussion with Attendees

SESSION VIII: "BLUE SKY" SESSION

Session Objectives: Explore options to bring about a paradigm shift to the public health countermeasures enterprise from research, development, and approval. Discuss strategies to implement change, including legal, statutory, and regulatory authorities.

2:40 p.m. Session Objectives and Introduction

GAIL CASSELL, *Workshop Chair*
Vice President, Scientific Affairs and Distinguished Lilly Research Scholar for Infectious Diseases
Eli Lilly and Company

2:45 p.m. Panel Discussion: Proposing Paradigm Shifts

GILLIAN WOOLLETT
Chief Scientist
Engel & Novitt, LLP

FRANK GOTTRON
Specialist in Science and Technology Policy
Congressional Research Service
U.S. Library of Congress

THOMAS MONATH
Partner
Pandemic and Biodefense Fund
Kleiner Perkins Caufield & Byers

BRETT GIROIR
Vice Chancellor for Research
Texas A&M

ERIC ROSE
CEO and Chair, Board of Directors
Siga Technologies, Inc.

BRUCE ARTIM
Director
Federal Affairs
Eli Lilly and Company

CHUCK LUDLAM
Former Counsel, Senator Joseph Lieberman
Former Principal Lobbyist for The
Biotechnology Industry Organization

PHILLIP GOMEZ
Director
Biodefense and Public Health Practice
PRTM

3:40 p.m. Discussion with Attendees: What Options Rise to the Top? What Have We Not Considered?

GAIL CASSELL, *Workshop Chair*
Vice President, Scientific Affairs and Distinguished Lilly Research Scholar for Infectious Diseases
Eli Lilly and Company

4:30 p.m. ADJOURN

C

Registered Workshop Attendees

Terry Adirim
Department of Homeland Security

Will Alameida
PanThera Biopharma

Jennifer Alton
Bavarian Nordic

Margaret Anderson
FasterCures

Lida Anestidou
The National Academies

Marietta Anthony
Critical Path Institute

Stacy Arnesen
National Library of Medicine/National Institutes of Health

Damon Arnold
Illinois Department of Health

Phyllis Arthur
Biotechnology Industry Organization, Inc.

Bruce Artim
Eli Lilly and Company

Guillermo Aviles-Mendoza
HHS/ASPR

Sheila Avruch
Government Accountability Office

Gerrit Bakker
ASTHO

Tali Bar-Shalom
Government Accountability Office

Cindy Bascetta
Government Accountability Office

Alison L. Bawden
Soligenix, Inc.

Ketty Belizaire
TB Alliance

Georges Benjamin
American Public Health Association

Kavita Berger
AAAS Center for Science, Technology and Security Policy

Kathleen Berst
CBMS-JVAP

Gregory Bice
ASDHA/FHP&R/DoD

Gary Bignami
PanThera Biopharma, LLC

Peter Blain
Health Protection Agency, UK

James Blumenstock
Association of State and Territorial Officials

Douglas Boenning
HHS

Brent Bomer
XOMA

Luciana Borio
FDA

Katie Brewer
American Nurses Association

Robert N. Brey
Soligenix, Inc.

Kathryn Brinsfield
Department of Homeland Security

Steve Brozak
StormBio Inc.

Greg Burel
Centers for Disease Control and Prevention/OPHPR

Stephen V. Cantrill
National Biodefense Science Board

Roberta Carlin
American Association on Health and Disability

Gail Cassell
Eli Lilly and Company

Stacey Chambers
HHS/NIH/NINDS

D. W. Chen
Department of Defense

Kelly Childress
Foley Hoag, LLP

Kenneth Cole
ODATSD (CBD/CD)/DoD

Susan Coller-Monarez
DHS

Chris Colwell
Becton Dickinson

Doug Compton
PhRMA

Francesca Cook
PharmAthene

Mark Cooke
PharmAthene

Susan Cooper
Tennessee Department of Health

Brian Corrigan
Astellas Pharma US, Inc.

Allan Coukell
The Pew Charitable Trusts

Brooke Courtney
Center for Biosecurity

Dack Dalrymple
Dalrymple & Associates, LLC

Christopher Davis
HHS

Anne De Groot
EpiVax

Walla Dempsey
HHS/NIH/NIAID

Gary Disbrow
HHS/BioMedical Advanced Research and Development Authority

Jacqueline Dombroski
AVI BioPharma, Inc.

Michael Doney
Centers for Disease Control and Prevention

Matthew Duchars

Lydia Duckworth
MITRE/HHS

Melissa Dupree
Fabiani & Company

Diana Elkis
PRTM

Peter Emanuel
OSTP, White House

David Evans
U.S. Army Medical Research & Materiel Command

Daniel Fagbuyi
Children's National Medical Center

Ann Farese
University of Maryland

Anthony Fauci
NIAID

Stuart Feldman
Sanofi Pasteur

Philip Ferro
HHS/ASPR

Patrick Flavin
Advanced Life Sciences, Inc.

Erin Fowler
HHS

Thomas Frieden
Centers for Disease Control and Prevention

Chuck Fromer
Chemical Biological Defense Directorate

Thomas Fuerst
HHS/ASPR

Nick Fullenkamp
Emergent BioSolutions

Claudia Gaffey
FDA

Alexander Garza
Department of Homeland Security

Bruce Gellin
HHS

Jeanne Geyer
FDA

Brent Gibson
Atlas Research, LLC

David Gilbert
Providence Health & Services

Brett Giroir
Texas A&M

Lewis Goldfrank
New York University School of Medicine

Mark Goldberger
Abbott Labs

Michael Goldblatt
Functional Genetics

Phillip Gomez
PRTM

Jesse Goodman
FDA

Frank Gottron
Congressional Research Service, U.S. Library of Congress

John Grabenstein
Merck & Co., Inc.

Dwayne Grant
HHS/OIG

Gigi Gronvall
Center for Biosecurity of UPMC

Robert Guidos
Infectious Diseases Society of America

Elin Gursky
ANSER

James Guyton
PRTM Management Consultants

Jami Haberl
Safeguard Iowa Partnership

Jeff Hackman
Intercell

Nate Hafer
NIH/OD

Marlene Haffner
Haffner Associates, LLC

Veronica Hall
Emergent BioSolutions

Michael Handrigan
HHS/ASPR

Julie Hantman
HHS/ASPR/BARDA

Maureen Hardwick
Alliance for Biosecurity

Chandresh Harjivan
PRTM

Richard Hatchett
White House, Executive Office of the President

Herbert Hecht
Emergent BioSolutions

William Helming
PRTM Management Consultants

D. A. Henderson
Center for Biosecurity of UPMC

David G. Henry
National Governors Association

Jack Herrmann
National Association of County and City Health Officials

Penny Hitchcock
Tauri Group

Sally Hojvat
OIVD/CDRH/FDA

Mary Homer
HHS/ASPR

Omar Hottenstein
Office of the Assistant Secretary of Defense for Health Affairs

Robert House
DynPort Vaccine Company, LLC

Jay Hunter
University of Maryland School of Medicine

Jean Hu-Primmer
FDA/Office of Counterterrorism and Emerging Threats

Patrick Iversen
AVI Biopharma

Rick Jaffe
ANSER, Inc., in support of ODATSD (CBD/CD)

James James
American Medical Association

Paul Jarris
Association of State and Territorial Health Officials

Daniel Jernigan
Centers for Disease Control and Prevention

David Jett
NIH/NINDS

Kelly Johnson
Agency for Healthcare Research and Quality

Vicki Johnson
National Commission on Children and Disasters

Jerry Johnston
National Association of Emergency Medical Technicians

Til Jolly
U.S. Department of Homeland Security, Office of Health Affairs

Robert Kadlec
PRTM Management Consultants

Brian Kamoie
White House, Executive Office of the President

Lisa Kaplowitz
Alexandria Health Department

James Kenimer
Biologics Consulting Group

John Kenten
Meso Scale Diagnostics

Thomas Keuer
ARCA biopharma, Inc.

Rima Khabbaz
Centers for Disease Control and Prevention

Lynne Kidder
Business Executives for National Security

Hye-Joo Kim
Centers for Disease Control and Prevention/Regulatory Affairs

Lisa Koonin
Centers for Disease Control and Prevention

George Korch
HHS

Akhila Kosaraju
SIGA

Nik Koterski
U.S. Army Medical Research & Material Command

Gerald Kovacs
HHS/BioMedical Advanced Research and Development Authority

Jon Krohmer
Department of Homeland Security

Michael Kurilla
National Institute of Allergy and Infectious Diseases

Laura Kwinn
HHS

Anne Laffoon
Government Accountability Office

Randy Larsen
Commission on the Prevention of Weapons of Mass Destruction Proliferation and Terrorism

Michael Latham
Centers for Disease Control and Prevention

Catherine Laughlin
DMID, NIAID

James Lawler
White House, Executive Office of the President

Matthew Lawlor
HHS/ASPR/BARDA

Theresa Lawrence
HHS

Eva Lee
Center for Operations Research in Medicine and HealthCare, School of Industrial and Systems Engineering,
Georgia Tech University

Janet Leigh
LSU Health Sciences Center

Brad Leissa
FDA/CDER/OCTEC

Rachel Levinson
Office of the Vice President for Research and Economic Affairs, Arizona State University

Deborah Levy
Division of Healthcare Quality Promotion, CDC

Thomas Lilburn
Bayer Healthcare, LLC

Carol Linden
HHS/BARDA

Alicia Livinski
National Library of Medicine, National Institutes of Health

Cynthia Love
National Library of Medicine, National Institutes of Health

Daniel Lucey
Georgetown University Medical Center

Charles Ludlam
Former Counsel, Senator Joseph Lieberman
Former Principal Lobbyist for The Biotechnology Industry Organization

Nicole Lurie
HHS/ASPR

Boris Lushniak
FDA

Jayne Lux
National Business Group on Health

Brian Lynch
BD Medical

Thomas MacVittie
University of Maryland, School of Medicine

Carmen Maher
FDA

Bert Maidment
NIH/NIAID

Michael Mair
HHS

Don Malinowski
HHS/ASPR/NBSB Staff

Monique Mansoura
HHS/ASPR/BARDA

Maria Marinissen
HHS/ASPR

Richard Martin
American Society for Radiation Oncology

Tyler Martin
Dynavax Technologies

James Matthews
Sanofi Pasteur

Maritza McIntyre
Bavarian Nordic

Teri Melese
UCSF School of Medicine

Raina Merchant
HHS/ASPR

Toby Merlin
Centers for Disease Control and Prevention

Jeremy Middleton
Elusys Therapeutics

Aubrey Miller
FDA

Georgianne Mitchell
City of Baltimore OEM

Thomas Monath
Kleiner Perkins Caufield & Byers

Melinda Moree
BIO Ventures for Global Health

Chanton Morel
London School of Economics

Carolina Morgan
Government Accountability Office

Erin Mullen
Rx Response

Rick Murray
American College of Emergency Physicians

Jomana Musmar
HHS/ASPR

Padma Natarajan
Infectious Diseases Society of America

Thomas Neal
MITRE

David Noll
Bavarian Nordic

Jeanne Novak
CBR International Corp.

Nancy Nurthen
Chemical Biological Defense Directorate

Michael Ochs
Infectious Diseases Society of America

Anthony Oliver
NHTSA, Office of Emergency Medical Services

Paul Owens
Lilly Research Laboratories

Christopher Ozeroff
ARCA biopharma, Inc.

George Painter
Chimerix, Inc.

Jennifer Pakiam
National Institutes of Health

Peter Palese
Mount Sinai School of Medicine, New York

Gerald Parker
HHS

John Parker
HHS

Tracy Parker
Department of Homeland Security

Peter Patriarca
Biologics Consulting Group

Andrew Pavia
University of Utah School of Medicine

Raquel Peat
FDA/CDRH/OIVD/DMD

Mary Pendergast
ARCA biopharma, Inc.

Dana Perkins
HHS/ASPR

David Perryman
Zirus, Inc.

Del Persinger
PhRMA Foundation

Cheryl Peterson
American Nurses Association

Sally Phillips
Agency for Healthcare Research and Quality

William Pinard
American Association for the Advancement of Science

Douglas Pon
Pfizer, Inc.

Elizabeth Posillico
Elusys Therapeutics

Beena Puri
FDA/CDRH/OIVD/DMD

Resha Putzrath
Navy and Marine Corps Public Health Center

Beth Rada
XOMA

Narayani Ramakrishnan
National Institute of Allergy and Infectious Diseases

Jacqueline Rams
Southern Research Institute

Stephen Redd
Centers for Disease Control and Prevention

Christopher Revere
National Commission on Children and Disasters

John Rex
AstraZeneca

Andrew Rickles
HHS/ASPR

David Ridley
Duke University

Corinne M. Ringholz
Chemical Biological Defense Directorate

Rosemary Roberts
Center for Drug Evaluation and Research, FDA

MacKenzie Robertson
HHS/ASPR

Robin Robinson
HHS

Michelle Rodrigues
SRI International

Paige Rogers
HHS/ASPR/BARDA

Eric Rose
SIGA Technologies, Inc

Philip Russell
Sabin Vaccine Institute

Jerald Sadoff
AERAS Global TB Vaccine Foundation

Elizabeth Sadove
FDA

Ronald Saldarini
Biological Initiatives

Emad Samad
StormBio

Tiffany Sampson
Department of Homeland Security

Leigh Sawyer
HHS

Patrick Scannon
XOMA

Julie Schafer
Biomedical Advanced Research and Development Authority

David Schmickel
FoxKiser

Roslyne Schulman
American Hospital Association

Talisha Searcy
HHS/OIG

Frederic Selck
Center for Biosecurity

Jonathan Shapiro
Department of Health and Human Services

Jerry Sheehan
National Library of Medicine National Institutes of Health

Marilee Shelton-Davenport
National Research Council/ National Academy of Sciences

Angela Shen
National Vaccine Program Office, HHS

Thomas Shipley
Department of Homeland Security

Roxanne Shively
HHS/ASPR/BARDA

David Siegel
NIH, NICHD

Michael Skidmore
Department of Defense

Laura Smart
Chimerix

Brad Smith
Center for Biosecurity, UPMC

Frank Sokolik
The Joint Commission

Daniel Sosin
Centers for Disease Control and Prevention

Raul Sotomayor
HHS/ASPR

Sharon Stanley
American Red Cross, National Headquarters

Craig Steinman
Abbott

Tim Stephens
National Sheriffs Association

Carolyn Stevens
HHS

Tom Stewart
AVI BioPharma

Deborah D. Stine
OSTP

Brook Stone
HHS/ASPR/NBSB

Raymond Strikas
National Vaccine Program Office, HHS

Victoria Sutton
Texas Tech University School of Law

Ernie Takafuji
National Institute of Allergy and Infectious Diseases

Carmen Tamayo
Foresight Links/FIP

Janet Tobias
Ikana Media

Eric S. Toner
Center for Biosecurity, University of Pittsburgh Medical Center

John Trizzino
Novavax, Inc.

Jessica Tucker
HHS

Craig Vanderwagen
HHS/ASPR (retired)

Rajeev Venkayya
Bill & Melinda Gates Foundation

Keith Vesely
Office of the Assistant Secretary of Defense for Health Affairs

Bruce Walker
Northrop Grumman Corporation

Cole Werble
The RPM Report

Gamunu Wijetunge
Department of Transportation

Ladd Wiley
Akin Gump Strauss Hauer and Feld

Andrew Willman
National Association of County and City Health Officials

Theresa Wizemann
Wizemann Scientific Communications, LLC

April Wood
American Red Cross

Gillian Woollett
Engel & Novitt, LLP

Raymond Woosley
Critical Path Institute

Casey Wright
HHS/ASPR

David Wright
PharmAthene

Justin Wright
BD Medical

David Wurtman
NexBio, Inc.

Debra Yeskey
HHS/ASPR/BARDA

Kevin Yeskey
HHS

David Yeung
NIH/NINDS

Lauren Yeung
FDA

Wesley Yin
Boston University

Yon Yu
Centers for Disease Control and Prevention

Stephanie Zaza
Centers for Disease Control and Prevention

D

Case Studies of HHS Chemical, Biological, Radiological, and Nuclear Medical Countermeasure Development Programs, Executive Summary

The following is a white paper prepared for the February 22–24, 2010, workshop on the public health and medical countermeasure enterprise, hosted by the Institute of Medicine Forum on Medical and Public Health Preparedness for Catastrophic Events and Forum on Drug Discovery, Development, and Translation. All opinions expressed in this paper are those of the author and not necessarily of the Institute of Medicine.

By Robert Kadlec
Vice President
Global Public Sector
PRTM

The US government has a long and complex history of medical countermeasure development programs for chemical, biological, radiological and nuclear threats. The goal of this document is to revisit those experiences in order to glean insights into overarching challenges, successful strategies, and areas for improvement across the mission space. Seven medical countermeasure programs were analyzed to identify the most significant contributors to risk for each program, and the factors that led each program to meet or fall short of its goals. Throughout the history of the mission, the US government has actively sought to lower barriers to industry participation and the "rules of the game" have been continually amended and improved. As such, each medical countermeasure program was subject to an evolving framework of rules, challenges, and opportunities.

The following seven medical countermeasure programs were examined for this analysis:

1. BIOTHRAX™ Anthrax Vaccine Adsorbed (AVA)
2. Recombinant Protective Antigen (rPA) Vaccine
3. ACAM2000™ Smallpox Vaccine
4. IMVAMUNE™ Modified Vaccinia Ankara (MVA) Smallpox Vaccine
5. Medical Countermeasures for Hematopoietic Acute Radiation Syn-drome (hARS)
6. Medical Countermeasures for Viral Hemorrhagic Fevers (VHF)
7. Broad Spectrum Antibiotics for Bacterial Threats

This executive summary compiles the key elements of each program across the spectrum of medical countermeasure activities. The key risk factors that impacted each program are summarized. Overarching conclusions from the case studies are drawn, and potential areas for future solutions are outlined.

CASE STUDY SUMMARY TABLES

The following tables are intended to depict the entire value chain for medical countermeasure development, from initial threat assessment and requirements determination to response and recovery activities. These tables demonstrate the dependencies throughout the operational spectrum for each medical countermeasure analyzed. The color coding assesses the relative risk associated with each medical countermeasure for each component of the framework. While the focus of this study is the advanced development and procurement of medical countermeasure, the risks of failure do not end when a product enters into the SNS but ultimately reside on whether that medical countermeasure can be distributed and administered to those who need it when they need it.

Case Study Summary: Left Side/Preparedness Activities Preparedness Assessment Framework Mission Components

Legend: Low Risk, Medium Risk, Medium/High Risk, High Risk, N/A	Threat Assessment	Requirements Determination	Product Development	Infrastructure	Acquisition	Stockpiling
AVA	Pre-DHS material threat determination	(-) Late formal determination of # doses needed (+) 2004 IOM Study on safety and efficacy	(+) Licensed product (+) Recent FDA approval for "3/3" and IM (-) Risk communication issues	(-) Limited production capacity (-) Historic FDA compliance issues	(-) *Ad hoc* acquisition (-) Perception of lobbying by company	(+) Longer shelf life
rPA	(-) Single threat scenario & methodology for risk assessment	(-) Dose requirement determined by policy process	(+) Decade of DoD research (-) Stability and consistency issues (-) insufficient # candidates	(-) Limited manufacturing capacity and expertise	(-) Commitment to procure while still in preclinical development	(-) Limited shelf life
ACAM2000	(-) No formal threat assessment	(+) One dose per every American 155 m ACAM2000	(+) Product improvement (+) Existing data increased confidence	(+) Leveraged Baxter production capacity and expertise	(-) One time buy of 48 m doses	(+) Awarding of 10-year warm base contract
MVA	(-) No formal threat assessment	(-) Ambiguity concerning recipients of MVA	(+) Milestone payments (-) Unprepared for normal changes in regulatory data requirements	(-) Foreign manufacturer	(+) PREP Act addresses company liability issues	(-) Uncertainty to warm base manufacturing
hARS	(-) Perceived threat has not necessarily parlayed into priority and resources	(-) Need to have clear requirements	candidates with commercial market and later in clinical development (-) Low funding and lack of clear regulatory pathways to evaluate efficacy constrain number of candidates	(+) Commercial market ensures manufacturing base (+) animal models development for hARS	(-) Two failed RFPs (+) Acquisition of commercial production for SNS	(-) Uncertain product characteristics (e.g., shelf life)
Antibiotics	(+) MTD for anthrax, plague, tularemia, ganders, meliodosis rickettsia	(-) Lack of clarity in the kinds and classes of antibiotics to purchase	(-) Seeking indications for existing products and funding novel approaches	(+) Existing commercial manufacturing base	(-) Uncertainty as to the type and kind to acquire	(-) High cost of stockpiling and sustainment
VHF	(-) Identified three specific threats: Ebola, Marburg, and Junin	(-) Draft requirements	(-) Novel candidate products (-) New technologies	(-) Limited BSL-4 testing capacity (-) No manufacturing capacity	(-) Acquisition strategy	(-) No stockpiling strategy

Case Study Summary: Right Side/Response Activities Preparedness Assessment Framework Mission Components

Legend: Low Risk, Medium Risk, Medium/High Risk, High Risk, N/A	Detect	Decide and Deploy	Dispense	Healthcare Response	Event-Related Response	Federal Planning	Recovery Phase Activities
AVA	(-) Limited diagnostics slow detection	(-) Delays in environmental/ clinical detection	(-) Delays in distribution and administration	(-) Limited/no hospital surge capacity	(-) State local preparedness for bioterrorism	(-) No USG Anthrax CONOPS	(-) Unapproved PEP indications (-) National vaccine contingency plan
rPA	(-) Limited diagnostics slow detection	(-) Delays in environmental/ clinical detection	(-) Delays in distribution and administration	(-) Limited/no hospital surge capacity	(-) State local preparedness for bioterrorism	(-) No USG Anthrax CONOPS	(-) Uncertain PEP properties
ACAM2000	(-) Limited environmental/ clinical detection	(-) Delays in environmental/ clinical detection	(+) Plan for prioritized use of multiple products	(-) Limited isolation bed capacity	(-) State local preparedness for bioterrorism	(+) USG smallpox response plan	(+) Established value of LA smallpox vaccine for PEP
MVA	(-) Limited environmental/ clinical detection	(-) Delays in environmental/ clinical detection	(-) How would determination be made to receive MVA	(-) Limited isolation bed capacity	(-) State local preparedness for bioterrorism	(+) USG smallpox response plan	(-) Uncertain value of MVA in PEP
hARS	(-) Lack of clinical diagnostics (biodosimetry)	(-) Delays in environmental/ clinical detection	(-) Uncertain supply chain integrity post event	(-) Limited/no hospital surge capacity	(-) State local preparedness for bioterrorism	(+) No complete USG CONOPS for medical response to IND/RDD	
Antibiotics	(-) Limited diagnostics slow detection	(-)Delays in environmental/ clinical detection	(-) Complex supply chain management	(-) Limited/no hospital surge capacity	(-) State local preparedness for bioterrorism	No USG CONOPS	(-) Likely compliance of long-term PEP with antibiotics
VHF	(-) Limited environmental/ clinical detection	(-)Delays in environmental/ clinical detection	(-) Complex supply chain management	(-) Limited isolation bed capacity	(-) State local preparedness for bioterrorism	No USG CONOPS for VHF	

The Elements of Risk

Scientific/Technical Risks

Assessing the risks of commercial drug development is complicated by the source of the candidate product, therapeutic class, and product type. There are generally higher clinical approval success rates for in-licensed candidates as compared to candidates that are self-originated. In-licensed drugs may benefit from screening or testing prior to licensing, and many have been acquired after clinical testing has begun.[1] Moreover, some argue that in-licensed candidates are subject to a more rigorous analysis than products that have been self-originated.

Success rates for new drugs also vary by therapeutic class.[2,3] The table below illustrates the variability between three classes of compounds.

	Probability of Licensure[4,5]	
Therapeutic Class	From Discovery	From Phase III
Systemic Anti-Infective	15%	79%
Anti-Neoplastic/Immunologic	7.1%	55%
Central Nervous System	3.8%	46%

These probabilities emphasize the dramatic increases in risk for early-stage candidates as compared to products later in advanced development. However, even when clinical safety studies have been completed success is far from guaranteed. The findings also suggest that product development programs and associated pipeline strategies must be tailored for the desired product category. While the anti-infective statistics above provide some optimism for medical countermeasures for biodefense indications, the many challenges of working with these threat agents limit any real ability to compare across categories of threat agents. Additional differences in the data were noted when comparing small and large molecules, such as monoclonal antibodies.[6]

[1] DiMasi, J., L. Feldman, A. Seckler, and A. Wilson. 2010 (Feb. 3).Trends in risks associated with new drug development: Success rates for investigational drugs. *Clinical Pharmacology and Therapeutics*. p. 3.
[2] Kola, I., and J. Landis. 2004 (Aug.) Can the pharmaceutical industry reduce attrition rates? *Nature Reviews: Drug Discovery* 3:711-715.
[3] DiMasi et al., 2010, p. 3.
[4] Ibid, p. 4.
[5] Ibid, p. 4.
[6] Ibid, p. 6.

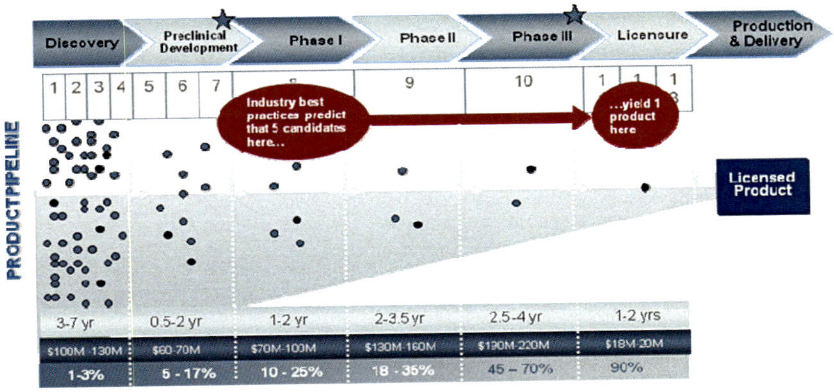

Probability of Success to Licensure

Numerous technical factors impact the chance of success of a product. Failure of large pharmaceutical company commercial drug trials are principally attributed to lack of efficacy (approximately 30%), adverse safety and toxicity profiles (an additional 30%), as well as commercial considerations (cost of development and estimated return on investments).[7] While failure is undesirable, the pharmaceutical industry views it part of the cost of doing business. Companies prefer when a candidate fails early in its development, as it limits actual and opportunity costs. For the development of CBRN medical countermeasures, failure must be viewed similarly: inevitable, acceptable, and ideally occurring at an early stage in development.

The lack of validated animal models emerged as a significant and recurring contributor to risk across all case studies. US government partners can do much more to coordinate across organizations, ensuring appropriate data sharing agreements, efficient validation processes, and more rapid means of finalizing animal model reports.

The main tool to mitigate scientific and technical risks is in building from existing products and product candidates—pursuing dual-use of commercially available products, or improving the characteristics of an already available medical countermeasure. Any efforts to move away from developing novel single-purpose products will help to reduce the research and development risks. The use of flexible platforms for development and production also represent a promising mechanism for reducing risk throughout the lifecycle of a product.

Legislative/Legal Risks

Project BioShield procurements have strict requirements on approval and contracting mechanisms that severely restrict the ability for the US government

---

[7] Kola and Landis, 2004.

to offer the most appropriate contracting vehicle to each performer. These acquisitions were given greater flexibility through the use of milestone payments authorized in the milestone funding that was enabled by the legislation contained in the Pandemic and All-Hazards Preparedness Act of 2006. BARDA has not yet utilized the Other Transaction Authority that has been granted to it, but NIAID has shown some success in using MTAs, CDAs, and modified clinical trial agreements to effectively collaborate with private sector partners.

The liability protections enacted in the PREP Act[8] has assuaged the concerns of many firms that develop medical countermeasures. This legislation mandates that manufacturers have total immunity from liability surrounding use of their products unless there is evidence of willful misconduct on the part of the company.

Financial Risks

The commercial pharmaceutical research and development of FDA approved vaccines, drugs and diagnostics are acknowledged as expensive and long processes with a substantial risk for failure. The current costs of bringing a new medicine from discovery through licensure are estimated at between $570 million to $1.7 billion.[9,10] In addition to the inherent costs, data derived from surveys of the major pharmaceutical companies indicate significant attrition. The current commercial trend indicates that 13 preclinical candidates are necessary to yield a single successful approved product—an approximately 8% chance of success.[11] For those candidates entering phase I clinical trials, only one in six (17%) make it to the marketplace.[12] Once a candidate enters Phase III clinical trials, it may only have a 64% chance of becoming an approved product.[13] Failures in Phase III are particularly expensive in both real and opportunity costs.[14] Finally, the time to identify and develop a new candidate vaccine or drug can be between seven to twelve years. The pre-clinical drug

---

[8] See https://www.safetyact.gov/jsp/refdoc/samsRefDocView.do?action=View Attachment&refDocGroupName=Reference ocuments&refDocTitle=Final SAFETY Act Rule (PDF)&attachmentName=Final SAFETY Act Rule.pdf.

[9] DiMasi et al., 2007.

[10] Tufts Center for the Study of Drug Development. 2001 (Nov.). *Backgrounder: How new drugs move through the development and approval process.* Boston. Gilbert, J., P. Henske, and A. Singh. 2003. Rebuilding big pharma's business Model: In vivo, the business & medicine report. *Windhover Information* 21:10.

[11] Hu, M., K. Schultz, J. Sheu, and D. Tschopp. 2007 (March 12). *The innovation gap in pharmaceutical drug discovery & new models for R&D success.* Chicago, IL: Kellogg School of Management.

[12] DiMasi et al., 2010, pp. 1-6.

[13] Ibid, pp. 1-6.

[14] Kola, I. 2008. The state of innovation of drug development. *Clinical Pharmacology & Therapeutics* 83; 227-230

discovery process can last from one to five years and the average time between the authorization of an Investigational New Drug (IND) entering into phase I clinical trials to the finalizing of the New Drug Application (NDA) or Biologics License Application (BLA) is approximately 6.4 years.[15]

The small number of successful CBRN medical countermeasure compounds the above industry percentages and adds significantly greater difficulty in attracting industry partners. In addition to the direct costs, commercial product developers potentially miss out on successful commercial products by investing time and capital into this niche market. Therefore, the opportunity costs alone limit the number of industry partners interested in initiating CBRN medical countermeasure programs.

Companies involved in the CBRN medical countermeasure market may also encounter unexpected financial challenges that require additional assistance from the USG. In addition, companies in this sector typically have limited financial resilience; without advance payments or milestone funding, typical technical or regulatory setbacks can threaten the viability of these companies. On the other hand, full US government support of single product manufacturers is an expensive venture. One developer of one product can easily require billions of dollars of investments for product development and licensure, warm-base manufacturing, stockpiling and replenishment activities.

Dramatic differences in portfolio spending have been noted, even among the top priority medical countermeasure programs highlighted in the *2007 HHS PHEMCE Implementation Plan*. Medical countermeasures for radiological and nuclear threats have received a small fraction of the funding allocated for anthrax medical countermeasures. While the radiological/threat medical countermeasure community has responded with increased collaboration and exploration of alternative development and stockpiling strategies, this lack of investment at all stages of the development pipeline continues to discourage most potential industry partners.

Regulatory Risks

Uncertainty about regulatory endpoints (definition of "usable product") and unanswered regulatory related science issues create further challenges. Technical limitations and significant scientific gaps increase the challenges of navigating the regulatory pathway. As animal models mature along with candidate products, additional knowledge that is uncovered during the process can significantly impact the pathway to licensure or approval of a product. Performers need to be prepared for such uncertainty.

---

[15] Berndt, E., A. Gottschalk, and M. Strobeck. 2005 (April 19). *Opportunities for improving the drug development process: Results from a survey of industry and the FDA,* presented at the National Bureau of Economic Research Workshop on Innovation Policy and the Economy, The National Press Club, Washington, DC.

The USG should examine ways to assist companies during the regulatory process. FDA has expended extraordinary energy in ushering inexperienced companies through the development process. Such companies are often characterized as requiring extensive "hand-holding" through assay development and product stability submission, and still encounter significant problems. Regulatory submissions are often of poor quality, with highly unrealistic timelines. These problems are compounded by the inability to attract large and experienced performers to this space.

Expanding indications for already existing products, such as broad-spectrum antibiotics, should be the most straightforward path for building the medical countermeasure stockpile. However, these efforts have been fraught with a lack of clear guidance and fractured government development processes. Interagency coordination and a clearly articulated goal are essential to success.

Organizational Risks

Besides the scientific and regulatory factors cited above, there are risk factors that are inherent to participating companies that affect the likelihood of successful CBRN medical countermeasure development. The quality of its management, the staff's technical experience, familiarity with the regulatory process, any past experience with successful licensure of a product, and the company's financial soundness affect the perceived likelihood of success. There are extrinsic factors that relate to the company's interface with the United States Government (USG), particularly through publication of medical countermeasure requirements and product characteristics, the Federal Acquisitions Regulations (FAR) contracting or research grant processes, and adherence to the FDA regulatory process. There are also policy and legislative considerations that play a role that are extrinsic to the company. Executive Branch and Congressional efforts to promote research and development of CBRN medical countermeasure reflect the desire by the USG to reduce the barriers to entry and offer incentives to increase the participation of companies.

There are other considerations as well. Perceptions of the CBRN medical countermeasures market, and hence profit potentials, have their own set of influences as to the kinds of companies and investors attracted to the CBRN medical countermeasure enterprise. To date, with few exceptions, large, experienced pharmaceutical companies have not entered the CBRN medical countermeasure market because of lack of sufficient profit or other incentives. This leaves the market to smaller and less experienced companies. The perception is that relying on companies, some without ever having successfully marketed a drug or vaccine, increases the risk of failure. Small to mid-sized companies are viewed as having limited financial reserves, inexperienced management, limited technical expertise, and limited experience with the FDA regulatory process. These numerous factors may have had synergistically negative effects on the outcome of a range of CBRN medical countermeasure programs.

Other Risks

Many medical countermeasures have been pursued without a clear vision of their effective use or acceptability to the public. Engagement with all stakeholders, from researchers and products developers to public health officials, medical providers, and the public is essential to an effective medical countermeasure enterprise. As an example, diagnostics are under-valued yet a crucial enabling capability that allow for more effective use of limited medical countermeasures. Without a clear understanding of utilization policies for diagnostics, informed by all necessary partners, diagnostics capabilities will never receive the high prioritization that they deserve.

In light of historic CBRN threats, the concern of terrorist acquisition of such weapons has lead to a sense of urgency. There is a perceived need to accelerate the development and production of CBRN medical countermeasure. According to the 2008 report by the Commission on Weapons of Mass Destruction (WMD), a WMD attack somewhere in the world, more likely biological than nuclear, is anticipated by 2013.[16] Their report mirrors internal USG assessments.[17] While this sense of urgency increases pressure, and expectations on these programs, there has not been a corresponding groundswell of public support or interest among US citizens. This results in programs that have substantial visibility but little support, and that require significant resources yet to date have low rates of return from the public's perspective.

The national security risks reflect the perceived catastrophic political, military and economic consequences that a large scale CBRN incident could have on the nation. For national security policy-makers, the greatest risk may be not having sufficient types and quantities of CBRN medical countermeasure when the situation demands. Given such high stakes, it is imperative that policy-makers fully understand the risks inherent to commercial drug or vaccine development, and support the appropriate resourcing and managing of such programs.

There is an association between the risks of medical countermeasure development and perceived national security risk. High risks associated with CBRN medical countermeasure development intuitively translate to greater risk from a national security perspective. However, the converse may not necessarily be true. Perceived high national security risks should not necessarily result in accepting greater technical risks of medical countermeasure development. On the contrary, it would seem essential, in light of national security risks, to mitigate the technical risks of medical countermeasure development and

---

[16] Graham, R., and J. Talent. 2008 (Dec.). *Nation at risk.* Weapons of Mass Destruction Commission Report.

[17] M. McConnell, Statement by the Director of National Intelligence, December 2008.

procurement. The objective should be to control and limit technical risks so as to modify the national security risks perceived. The risk calculations for CBRN medical countermeasure development are therefore complex, multi-factorial and cross private and public domains subject to factors with and without historical precedent. It is incumbent upon the USG to provide the necessary assistance to mitigate as many risks as possible.

US GOVERNMENT RESPONSES TO RISK

The USG has conducted research and development of CBRN medical countermeasures since World War II. The Department of Defense (DOD) has historically had the principal role in CBRN medical countermeasure development. DOD's requirements are to ensure the protection of its military force conducting assigned military operations. The preponderance of this research and development effort has been focused on vaccines and drugs for pre-exposure protection. Despite its long-standing efforts in this endeavor, external assessments determined that DOD was not organized to develop and license vaccines, therapeutic drugs, and antitoxins. Their efforts were characterized as disjointed, with the following factors cited: fragmentation of responsibility and authority, changing strategies, a lack of strong management, limited technical expertise, and a financial commitment that was not commensurate with the requirements of its program goals.[18,19]

In 1998, in response to the growing concern about the risk of WMD terrorism, President Clinton designated the Department of Health and Human Services (HHS) as the Lead Federal Agency in planning and preparing for response to domestic WMD-related medical emergencies.[20] HHS, in conjunction with the Department of Veterans Affairs, was directed to stockpile antidotes and pharmaceuticals in the event of a WMD incident. The National Pharmaceutical Stockpile (later renamed the SNS) was the first ever civilian medical stockpile, containing necessary medication to treat those exposed to biological or chemical weapons.[21] Contrary to DOD's pre-exposure approach, HHS research and development efforts focus primarily on post-exposure prophylaxis and therapeutics. In addition to its different strategic approach, HHS has to consider the greater diversity of potential recipients of these medical countermeasures, including special populations such as children, the elderly, and the immunocompromised.

---

[18] Institute of Medicine and National Research Council. 2004 *Giving full measure to countermeasures*. Washington, DC: The National Academies Press.
[19] Anna Johnson-Winegar Testimony, 2001.
[20] Presidential Decision Direction 62. 1998 (May 22). Combating Terrorism.
[21] The National Pharmaceutical Stockpile was renamed the Strategic National Stockpile in The Homeland Security Act of 2003.

Following the terrorist airline and anthrax letter attacks of 2001, the USG significantly increased its investments in basic research and development of CBRN medical countermeasures, as well as the acquisition of countermeasures, for the expansion the SNS. The USG has appropriated and spent approximately $55 billion for biodefense preparedness, including researching, developing, and acquiring CBRN medical countermeasures.[22,23] Approximately $1.6 B per year has been spent by the NIAID in basic research since FY 2004. Congress also appropriated $5.6 B for acquisition of CBRN medical countermeasures through Project BioShield.

President Bush issued several Homeland Security Presidential Directives (HSPD) pertaining to medical countermeasure development: HSPD-10, National Biodefense Policy; HSPD-18, WMD Countermeasures; and HSPD-21, Medical and Public Health Preparedness. These policy documents provided guidance pertaining to the importance, approach to research and development, and utilization of CBRN medical countermeasures. HPSD-18 states "the development and acquisition of effective medical countermeasures to mitigate illness, suffering, and death resulting from CBRN is central to our consequence management efforts." This document also states "it is the policy of the U.S. to draw upon the considerable potential of the scientific community in the public and private sectors to address our medical countermeasure requirements relating to CBRN threats." However, none of the policy documents specifically address the inherent challenges in either developing or acquiring such products. HSPD-18 simply states that creating such medical countermeasures is a "time-consuming and costly process."[24]

In the Public Health Security and Bioterrorism Preparedness and Response Act of 2002 (PL 107-188) Congress affirmed the priority of CBRN medical countermeasures by directing the Secretary of HHS to

- ensure development and acquisition of smallpox vaccine;
- accelerate the approval of priority medical countermeasures as a fast-track product through the FDA;
- direct the FDA to develop the Animal Rule; and
- accelerate research and development on medical countermeasures against top-priority threat agents.

Congress also played a significant role in incrementally lowering the perceived barriers to medical countermeasure development. The initial

---

[22] See http://www.armscontrolcenter.org/policy/biochem/articles/fy09_ biodefense_ funding.

[23] Franco, C. 2009. Billions for biodefense: Federal agency biodefense funding, FY2009-FY2010. *Biosecurity and Bioterrorism*: 7. 1-19.

[24] HSPD-18 Weapons of Mass Destruction Countermeasures, The White House, January 31, 2007.

legislative actions were directed to entice large pharmaceutical companies to participate in CBRN medical countermeasure development. The requisite measures to overcome perceived barriers to their participation were several: guaranteeing a medical countermeasure market,[25,26] limiting liability, improving USG contracting practices, clarifying regulatory guidance, and expediting the regulatory process.

President Bush proposed the Project BioShield Act in 2003, which attempted to address the need for a guaranteed USG market for CBRN medical countermeasures. The original proposal would have permitted payment for acquisition directly from the US Treasury, which would have circumvented the annual Congressional authorization and appropriation process. This approach would have provided unlimited funds for medical countermeasure acquisition that were deemed necessary and sufficiently mature for inclusion into the SNS. Avoiding the annual appropriations process and permitting unlimited expenditures was believed to be sufficient incentive to induce large, more experienced companies to enter the CBRN medical countermeasure market. As part of this initiative, an extensive oversight mechanism was created, requiring the Secretaries of HHS and Homeland Security (DHS) and the President to certify that the acquisition of the medical countermeasure was necessary.

Congress balked at such a proposition based on its Constitutional responsibilities for conducting oversight and providing regular appropriations. Congress eventually passed a bill in 2004 that created a $5.6 B discretionary reserve fund for procuring CBRN medical countermeasures for the SNS. Though the indefinite mandatory authorization was not passed, the extensive oversight mechanisms requiring DHS, HHS and Presidential concurrence were retained. In addition to these funds, the Act included provisions to increase NIH (NIAID) authorities and flexibility to expedite basic research and development of medical countermeasures; it also created the Emergency Use Authorization process for FDA.[27]

While BioShield created a guaranteed Federal market for CBRN, it was a limited one-time appropriation. Large established pharmaceutical companies were not attracted to the CBRN medical countermeasure market. Other significant barriers remained, such as product liability, that increased the perceived and potential risks to established companies. Since these medical countermeasures would not likely have extensive human testing or use prior to an emergency, companies feared that widespread use of their products in an emergency could result in previously unrecognized adverse events that the companies could be held liable for.

---

[25] Miller, H., and S. Kazman. 2002. Federalize vaccine production? We'd be taking a shot in the dark. *Hoover Digest*, No. 2. www.hoover.org/publications/digest/3437401.html.

[26] Johnson-Winegar Testimony, 2000.

[27] Project BioShield Act of 2004. PL 108-276.

Additionally, BioShield initially mandated that payment was conditional upon product delivery; the only exceptions were advance payments—capped at 10% of contract value—that could be made before delivery. BioShield contracts were fixed price contracts that were viewed by HHS and their industry partners as overly restrictive. In later legislation, Congress authorized HHS to award milestone payments of 5% at various points in medical countermeasure development, with cumulative awards limited to 50% of the total contract value.[28]

In 2005, the threat of a looming H5N1 influenza pandemic prompted Congress to pass the Public Readiness Emergency Preparedness Act (PL 109-148). The Act limits claims that result from injuries or death from public health medical countermeasures (for pandemic and CBRN threats) when used during a declared public health emergency, or if there is credible risk of such an emergency. It specifically protects manufacturers, distributors, program planners, persons who prescribe, administer, or dispense countermeasures, and employees of any of the above. It provides manufacturers absolute immunity, except if injury or death resulted from willful misconduct.[29] Limiting liability significantly affected the willingness of major seasonal influenza vaccine manufacturers to develop and manufacture pandemic influenza vaccines. In contrast, it did little to incentivize companies to participate in CBRN development.

The Pandemic All-Hazard Preparedness Act of 2006 (PL 109-417) specifically addressed the reality that large experienced companies generally have not entered the CRBN medical countermeasure market. Rather than seeking to create further incentives for large pharmaceutical manufacturers, the bill attempted to mitigate the perceived financial, contracting, legal, and regulatory risks of the medical countermeasure development process for the smaller, less experienced companies that were participating. The bill also granted HHS limited antitrust exemptions to facilitate collaboration between companies involved in medical countermeasure development.

The bills' sponsors, Senators Burr and Kennedy, proposed creating the Biomedical Advanced Research and Development Authority (BARDA) under the Assistant Secretary of Preparedness and Response (ASPR) to support, coordinate, and provide oversight of advanced development of pandemic vaccines and CBRN medical countermeasures. The act required that HHS facilitate and increase communication between the HHS, FDA and the participating companies, and establish strategic initiatives to accelerate development and promote innovation in CBRN medical countermeasure. It provided HHS with the authority to award contracts, grants, and cooperative agreements or other transactions to promote innovation, development of research tools and research into rapid diagnostics, broad spectrum antimicrobials, and vaccine manufacturing technologies. BARDA was granted

---

[28] Public Readiness Emergency Preparedness Act of 2006. PL 109-148.

[29] Division C of PL 109-148 (2005), 42 USC 247d-6d, 247d-6e.

the same authorities as DOD's Defense Advanced Research Project Agency (DARPA).

Over time, all these executive and legislative actions shaped and modified the medical countermeasure landscape. Each separate action attempted to clarify the strategic priorities of and perceived obstacles to the USG medical countermeasure effort. The cumulative result of these efforts is a loose coalition of Executive Branch agencies that attempt to align and leverage their activities in medical countermeasure development, acquisition, distribution, and utilization.

President Obama announced before Congress during his 2010 State of the Union Address the "launching a new initiative that will give us the capacity to respond faster and more effectively to bioterrorism or an infectious disease—a plan that will counter threats at home and strengthen public health abroad."[30] His commitment reinforces the earlier announcement by Secretary of HHS to evaluate the capacity and capabilities of the United States to develop and manufacture vaccines and other products against influenza pandemics and bioterrorism.[31]

The US Government now retains CBRN medical countermeasure assets valued at nearly $4 billion and has committed significant additional funding for infrastructure including biocontainment, manufacturing, animal models, public health and hospital preparedness. There have been successes in the HHS medical countermeasure enterprise, but also ample opportunity to make improvements in its efficiency and effectiveness. The definition of success for the development of CBRN medical countermeasures can be broadly characterized as having a licensed, safe and efficacious medical countermeasure that can be rapidly distributed and administered to mitigate sickness, injury, or death resulting from exposure to CBRN agents.

OVERARCHING CONCLUSIONS

1. The commercial research and development pathway for drugs and biologics is lengthy, expensive, and risky; the CBRN medical countermeasure pathway compounds these challenges with additional financial and regulatory risks.

Failures in the commercial drug development process are principally due to shortfalls in product efficacy, adverse safety and toxicity profiles, and undesirable returns on investment. The industry has developed sustainable

---

[30] President Barack Obama, Remarks by the President in State of the Union Address, U.S. Capitol, January 27, 2010, http://www.whitehouse.gov/the-press-office/remarks-president-state-union-address.

[31] The White House. *National Strategy for Countering Biological Threats*. By President Barack H. Obama. November 23, 2009. http://www.whitehouse.gov/sites/default/files/National_Strategy_for_Countering_BioThreats.pdf.

business models that account for substantial research and development costs and significant candidate attrition. The scientific, regulatory, and financial foundations of CBRN medical countermeasures are significantly more unstable than for mainstream drug and biologic products. In addition, the failure of a CBRN medical countermeasure candidate has been viewed as a catastrophic event by both government and industry participants; these failures need to be incorporated into the system as inevitable and acceptable risks of the business, as they are in mainstream pharmaceutical development. The goal should be to ensure that potential failures occur at the earliest possible stages of the product development process.

2. If considered solely on the basis of financial risk, most CBRN medical countermeasures would never be developed.
The national security consequences of a large scale CBRN attack compel the US government to lead the pursuit of safe and effective medical countermeasures to protect the United States against these threats. However, the current risks of medical countermeasure development are complex and cross private and public domains. It is essential that all partners work aggressively to mitigate potential risks throughout the development process. These risk categories include scientific/technical, organizational, regulatory, financial, and policy.

3. Despite some high-profile failures, the CBRN medical countermeasure development process has improved over time, and there are clear success stories to learn from. In responding to the threat of smallpox, HHS provided strong Secretarial leadership, created specific requirements for the desired product, and focused necessary resources to develop, manufacture, and sustain a smallpox vaccine capability to protect every United States citizen. Since the Project BioShield Act of 2004, successive legislative actions have also expanded authorities for contractual flexibility, and provided appropriations, as well as limit liability, and created long-term (warm-base manufacturing) contracts, and milestone payments. Many challenges remain in developing a sustainable medical countermeasure industry, but there are positive lessons to be drawn from past experience, and a wealth of tools that the US government can use to improve the process.

4. Senior US government officials must reaffirm the high priority of the CBRN medical countermeasure mission. Leadership from the highest levels is essential for motivating and focusing the wide range of stakeholders to successfully prepare for these threats, and for attracting a large and diverse population of product developers. A clear signal of a long-term commitment to this market is essential for investors and industry partners. The ongoing shortfalls in available resources for product development leads to the perception that these activities are unimportant. Pharmaceutical and biotechnology companies and the investment community know the costs of commercial drug and vaccine development. US government funding to date has not been commensurate with commercial industry standards or with the perceived high risk of these programs. Appropriate investments have not been appropriated or allocated for any one, much less all thirteen, of the identified material threat agents.

5. Increased clarity and transparency in product requirements and desired characteristics are essential for every CBRN medical countermeasure. The *2007 HHS PHEMCE Strategy and Implementation Plan for CBRN Threats* represented the first effort by the US government to project future CBRN medical countermeasure needs. However, this guidance was only at the level of product category, and provided no additional detail as to the desired characteristics or quantities of each class of product. Past ambiguities concerning USG priorities, commitment, and requirements have undermined the confidence of participating companies and their investors, and have limited the interest of potential industry participants.

6. The HHS oversight and management of CBRN medical countermeasure development has improved over time, but still does not reflect comprehensive end-to-end portfolio and product management. Participating companies large and small will always encounter challenges in developing these novel products as a result of the myriad scientific and technical hurdles. However, it is in the vital interest of the mission that the US government do everything possible to mitigate all foreseeable and avoidable risks. The US government management of programs has substantial room for improvement. Prioritization by senior US government officials, combined with extensive in-house product development expertise, is central to improving oversight and management. The oversight and management of medical countermeasure basic research, through advanced development to stockpiling and sustainment, remains fragmented across several different agencies and budgets. Relationships across departments and agencies must be strengthened and formalized. The basic research portfolio does not appear to be optimally aligned to support the top-priority medical countermeasures outlined in the *2007 HHS PHEMCE Implementation Plan*. A consequence of that suboptimal alignment is a far smaller number of candidate products than is necessary to counteract the expected rate of candidate attrition. At the other end of the pipeline, costs of sustaining and replenishing the Strategic National Stockpile (SNS) have not been factored into current and future budget estimates, highlighting the need for a planning and budgeting process longer than the annual appropriations cycle.

7. The regulatory process supporting CBRN medical countermeasure approval and licensure demands improvement. Incomplete and sometimes conflicting guidance to participating companies from HHS agencies has created confusion that has led to unnecessary or duplicative studies, and potentially wasted time and resources. As the contracting agencies, BARDA and NIAID offer companies regulatory advice that can be inconsistent or conflict with FDA guidance. A lack of regulatory expertise from industry developers only compounds these difficulties, as many companies are completely reliant on government partners for navigating the regulatory process. Improvements in the science that underpins the regulatory process is also essential. The understanding of the pathobiology of these diseases and the host responses to CBRN agents is lacking in many cases. Further clarity and guidance with respect to effective use of the animal rule is also necessary; most threat agents do not yet have validated animals models, or correlates of protection from animals to humans. Additionally, the FDA has been tasked with this substantial new mission since 2002, and has received little to no additional funding to support development of the necessary human capital to review submissions and provide guidance. At present FDA does not receive funding either through its annual budget or user fees to support its regulatory review of CBRN medical countermeasure products.

8. With few exceptions, experienced pharmaceutical companies have not entered the CBRN medical countermeasure market. Smaller and less experienced companies predominate, and present a variety of partnering challenges. Without expert assistance, vaccine and drug development by inexperienced companies is associated with a significant risk for failure. These companies generally have not successfully achieved FDA approval or licensure of any products, have limited technical expertise or experience with the FDA regulatory process, and require coordinated assistance from experts at BARDA, NIH and FDA.[32] Building a cadre of HHS experts in product development and manufacturing is critical to supporting this endeavor, and requires the government to place a high priority on developing an elite workforce.

In addition, the performers in the CBRN market generally have limited financial resilience and are almost completely dependent on external funding.[33] Their financial standing can be severely compromised when faced with additional studies and trials; more established performers would have sufficient resources to weather these anticipatable delays. The financial weakness of each separate performer is compounded into a significant vulnerability for the sustainment of the long-term medical countermeasure mission.

---

[32] Bolken, T., and D. Hruby. 2008. Discovery and development of antiviral drugs for biodefense: Experience of a small biotechnology company. *Antiviral Research* 77:4-5.

[33] Ibid, p. 4.

LESSONS LEARNED AND POTENTIAL SOLUTIONS

These case study analyses evaluate the perceived relative risks associated with the development and production of selected CBRN medical countermeasures. The risks are qualitatively weighed to reflect the preponderance of data collected through interviews, literature and official document reviews. It is worth noting that a risk resides in the eye of the beholder, and reflects the perspective of the observer. Participants in different components of the medical countermeasure enterprise will weigh relative risks very differently. Furthermore, the CBRN medical countermeasure development landscape is not static. Since 2000, the USG has actively sought to lower barriers to industry participation and the "rules of the game" have been continually amended and improved. As such, each medical countermeasure analyzed was subject to a changing set of rules, challenges, and opportunities.

Despite the process volatility, there are several practices that the USG has done right. For example, HHS created specific requirements, providing clear priority and strong leadership from the Secretary of HHS, and the necessary resources to develop, manufacture, and sustain a medical countermeasure capability against smallpox. The successive legislative actions—the 2002 Bioterrorism Preparedness, Project BioShield, PREP, and PAHPA Acts—all expanded authorities to increase contractual flexibility, increase appropriations, limit liability, create long-term (warm-base manufacturing) contracts, and introduce milestone payments. The summation of these actions incrementally improved the process by which the HHS pursues CBRN medical countermeasure development and procurement.

These successes can be replicated and built upon. However, the remaining challenges must be recognized and addressed. HHS can improve the likelihood of success for current and future programs by exploring some of the following approaches.

First, the perceived USG priority and need for CBRN medical countermeasures is inconsistent and unclear. The USG and the Secretary of HHS must reaffirm the priority and need for these products. Companies partially base their decisions on entering the medical countermeasure market on their determination of the USG's commitment and level of dedicated investment. Ambiguity on the part of the USG creates uncertainty for the companies and their investors. Beyond the strategic ambiguity, there is also ambiguity as it relates to the overall requirements and the prospective product characteristics that can result in unsatisfactory or incomplete Request for Proposals (RFP). The lack of specificity in RFPs can introduce potential candidates that do not meet the real needs or are at the wrong stage of product development for consideration. All these factors can dramatically extend the administrative time and cost of medical countermeasure development.

Changes in anticipated acquisitions, such as the cancellation of RFPs, undermine the confidence of companies and the investment community. The ambiguity in priorities is enhanced by a perception that the USG has devoted an unrealistically low level of funding to advanced development and stockpile sustainment of medical countermeasure. The investment community is well acquainted with the cost of commercial drug and vaccine development. Investors assess that the USG funding is not consistent with commercial industry standards and is not commensurate with the rhetoric concerning the perceived risk. Furthermore, the 13 published Material Threat Determinations (MTD) have not been prioritized. By neglecting to do so, the USG fails to convey a relative importance among the 13. HHS has not budgeted or appropriated sufficient advanced development funding for any one medical countermeasure program, much less for all medical countermeasures for all 13 identified material threats.

There is a significant level of concern regarding the long-term outlook for the existing USG funding for CBRN medical countermeasures. In the FY2010 budget request, the Obama administration approved transferring roughly $609 million from the Project BioShield Special Reserve Fund to NIAID ($304 million) and BARDA ($305 million).[34] While the support of additional advanced development activities is essential for success, removing the necessary resources from the only dedicated procurement fund sends mixed signals to potential developers.

After this transfer, the Project BioShield Special Reserve Fund has approximately $2.4 billion remaining through fiscal year 2013. This raises an even more significant question: what will the USG biodefense procurement budget be after 2013? Uncertainty surrounding the USG commitment to biodefense affects both company interest and investor confidence in this sector.

There are significant opportunities for improving the HHS organizational element of the medical countermeasure effort. The USG oversight and management of CBRN medical countermeasure development has evolved and improved but still does not reflect an "end to end" or "cradle to grave" comprehensive approach. When employed, this comprehensive approach has had tangible success; interviewees characterized development of ACAM2000™ as an "end to end" effort that resulted in successful stockpiling of a new smallpox vaccine.

Strong leadership and priority from senior USG officials, combined with extensive technical expertise in product development, is central to improving that oversight and management. Responsibilities for oversight and management of medical countermeasure basic research, through advanced development to stockpiling and sustainment, remain fragmented across several different agencies and budgets. The integration of efforts across participating agencies must be strengthened. The NIAID CBRN research portfolio does not appear to be optimally aligned to support the priorities of medical countermeasure development outlined in the *2007 HHS PHEMCE Implementation Plan for CBRN Threats*. A likely consequence of that suboptimal alignment is that there

---

[34] See http://www.globalsecuritynewswire.org/gsn/nw_20100108_2084.php.

are too few prospective CBRN medical countermeasure candidates to overcome the probabilities of failure for the development process. Furthermore, the costs of sustaining and replenishing the SNS have not been factored into current and future budget estimates; this highlights the need for a longer term planning and budgeting process than the annual appropriations cycle.

The regulatory process supporting the CBRN medical countermeasure development enterprise demands improvement. Incomplete and sometimes conflicting guidance to participating companies from HHS agencies has created confusion and may have increased costs. BARDA and NIAID offer companies regulatory related suggestions that can conflict with those that they receive from FDA. This guidance does not appear to be aligned across federal agencies or synchronized with the FDA.

Companies desire clear regulatory guidance to facilitate their product development. However, the regulatory science supporting CBRN medical countermeasure approval and rules for the development of CBRN medical countermeasure is not mature. The basic science underpinning CBRN medical countermeasure is an evolving field. As more knowledge and data is accrued, the ability of the FDA to provide regulatory guidance will likely improve. The animal rule requires further clarity and guidance, as there are not validated animal models for all the CBRN threat agents. Improving regulatory science is essential to improving the overall CBRN medical countermeasure enterprise. However, the FDA does not receive funding either through its annual budget or user fees to expand its scientific knowledge or support its regulatory review of CBRN medical countermeasure candidates.

Finally, the majority of medical countermeasure contracts (from BARDA and NIAID) are awarded to less experienced biotech companies, which represents a significant additional risk to successful product development. These companies generally have limited financial resilience and are dependent on external funding. They have limited assets, limited or non-existent revenue streams, and are heavily dependent on USG funding.[35] Their financial standing can be severely compromised when technical or regulatory issues are encountered that require additional studies or trials, thus increasing the costs of development. These companies may lack the financial resilience to enable survival through the product development process. They may require USG grants or subsidies to not only successfully develop a product, but also to become financially viable companies to ensure long-term sustainment of that product.

These companies also lack broad or in-depth in-house technical expertise and also lack the appropriate supporting infrastructure for manufacturing or testing and evaluation.[36] These companies require technical assistance that may or may not be available from BARDA, NIH, or FDA.[37] Similar to the limited

---

[35] Bolken and Hruby, 2008, pp. 4-5.
[36] Ibid.
[37] Ibid.

technical expertise, there appears to be limited experience with the regulatory and licensure process. The simple summation of these individual risks may not entirely reflect the cumulative risk that these companies face. The risk of less experienced companies are perceived to exceed the "ordinary and expected" risks associated with commercial drug and vaccine development by large, experienced pharmaceutical companies.

Commercial drug and vaccine development is challenging. It requires managing the scientific, technical, and regulatory risks to produce a profitable outcome. The US government's effort to successfully develop, procure, stockpile, and effectively use CBRN medical countermeasures is even more challenging. The pressing national security risks should compel all stakeholders to maximize resources, coordinate efficiently, and pursue all possible avenues to ensure that medical countermeasures are available to protect the public from the catastrophic outcomes of a potential CBRN event.

E

Synthesis of Business Models and Economic and Market Incentives for Vaccines and Therapeutics

The following is a white paper prepared for the February 22–24, 2010, workshop on the public health and medical countermeasure enterprise, hosted by the Institute of Medicine Forum on Medical and Public Health Preparedness for Catastrophic Events and Forum on Drug Discovery, Development, and Translation. All opinions expressed in this paper are those of the author and not necessarily of the Institute of Medicine.

By James Guyton, Hannah McClellan, and Fanni Li
Public Health and Biodefense Practice
PRTM

ABOUT THIS PAPER

Background and Charge

This paper has been developed to support an HHS Secretary-directed review of the Medical Countermeasures Enterprise (MCME) that addresses public health emergency threats including chemical, biological, radiological, and nuclear (CBRN) agents as well as pandemic influenza and other emerging infectious disease (EID). The objective is to conduct "a review of [HHS's] entire public health countermeasures enterprise ... to look at how our policies affect every step of countermeasure development and production and then ask: how can we do better?" The Secretary has charged the Office of the Assistant Secretary for Preparedness and Response (ASPR) with leading the review, given the ASPR's responsibility for directing and coordinating HHS's activities relating to protecting the U.S. population from acts of terrorism and other public health and medical threats and emergencies. This white paper will be used by the ASPR and by subsequent planning committees to develop public and stakeholder workshops to examine alternative methods and models for achieving successful product development, approval, procurement, and delivery to the U.S. populations.

As part of the larger public health enterprise review, this white paper explores the following topic: *Synthesis of Business Models and Economic and Market Incentives for Vaccines and Therapeutics*.

Paper Objectives

The objective of this paper is to explore alternative policies, business models, and incentives that can be used to foster a more effective and sustainable medical countermeasure enterprise. Particular focus will be placed on identifying ways to further the pharmaceutical industry's engagement in the MCME to move candidate medical countermeasures through advanced development and provide approved or licensed products for operational use. To this end, this paper will identify

1. challenges to engaging industry in the MCME and what is needed to overcome those challenges;
2. new and innovative policies, strategies, and incentives to encourage industry participation in the MCME; and
3. issues related to pursuing these new and innovative policies, strategies, and incentives.

Scope

This paper focuses primarily on policies, business models and incentives for increasing industry involvement in the MCME's programs for medical countermeasures for CBRN threats. The paper does not focus on the Pandemic Influenza program, given the already high level of involvement of multiple large-scale commercial vaccine manufacturers in the program, although the program is considered within the context of other models that may offer some approaches that could be applied to CBRN.

Methodology

The findings in this paper have been synthesized through a review of several literature sources, including published papers, MCME agency documents, and public presentations. This literature has been supplemented by findings from interviews with numerous stakeholders from industry, academia, and government agencies represented in the MCME. A bibliography and list of interviewees are provided at the end of the paper.

This paper has benefited from multiple interactions with participants in the February 23–24, 2010, Institute of Medicine workshop on The Public Health Emergency Medical Countermeasures Enterprise: Innovative Strategies to Enhance Products from Discovery through Approval as well as members of the National Biodefense Science Board's Markets and Sustainability Working Group.

EXECUTIVE SUMMARY

MCM Development Challenges

Developing medical countermeasures is critical to achieving the mission of protecting the U.S. population from acts of bioterrorism and other public health threats and emergencies. The MCM landscape is plagued by uncertainty in an already complex and challenging field where the development of pharmaceuticals and vaccines is inherently risky, lengthy, and costly. Successful achievement of mission goals will require close collaboration and partnership between the USG and public sector. Unfortunately, engaging experienced industry players, particularly large pharmaceutical companies, has proven challenging under the current MCME business model.

Current Approach to MCM Development

The Enterprise's CBRN program investments to date have primarily focused on biological threats. Policy decisions on how MCM products against biological threats will be used emphasize post-event response with stockpiled MCM products. To obtain these MCMs, the MCME is seeking to develop new products (vaccines and therapeutics) for a diverse set of requirements (including special needs populations) against thirteen Material Threat Determinations (MTDs). To fulfill this mission, the USG is partnering with MCM developers by employing a variety of incentives. Incentives include "push" mechanisms, such as grants and contracts for basic research and advanced development, as well as "pull" mechanisms, such as BioShield procurement contracts to entice MCM developers to develop MCMs through licensure or approval and produce them for procurement. To date, the incentives used to promote the MCME have succeeded in motivating significant engagement primarily by small innovator biotechnology companies.

Industry Needs

Incentives employed to date by the MCME are seen by industry to be insufficient to support robust development programs and to sustain a reliable market. As such, this approach has not created the conditions that would attract experienced industry participation. A more effective business model for MCM development could increase industry engagement by more successfully meeting the core needs of experienced pharmaceutical companies. To that end, we have identified three principal conditions that must be addressed:

- Product Requirements: Developers need specific requirements for the MCM, including *what* the product should be (Target Product Profiles [TPP], including formulation, dosage, method of administration, etc.), *how much* will be required, and *when* it must be delivered.
- Regulatory Clarity: Developers must have a clearly defined regulatory path to licensure, particularly with respect to the Animal Efficacy Rule requirements.
- Return on Investment: Companies must realize adequate returns, financial or otherwise, to offset the opportunity costs of other potential projects.

Medical Countermeasure Business Model Framework

Top-down course corrections will likely be necessary to resolve the observable disconnect between industry needs and the current approach to MCM development to better engage private sector partners. The challenge at hand will not be solved with a short term solution, but rather through a series of policies and strategies coupled with tactical practices and incentives that will enable the current "business model" to evolve.

The current approach to MCM development can be summarized as a business model composed of four basic components for achieving the goals of the MCME. The business model framework includes strategic, operational, and tactical planning elements, each of which plays a pivotal role in defining, organizing, and executing MCM development. The four elements of the MCME business model include to following:

- Policies for Product Use: Policy decisions for how MCMs are used (e.g., stockpiling, vaccination/prophylaxis that drive MCM product requirements (what, and how much).
- Product Strategies: Development of new MCM products vs. new indications for commercial products, or indications for special populations.
- Players and Roles: Company and customer types, roles in development, and structure of partnerships.
- Push and Pull Incentives: Incentives provided by the government to increase industry interest in MCM development opportunities.

Incentives

Incentives are a critical component of the business model, as they provide toolkits for executing strategic and operational plans. Incentives are generally grouped into "Push" incentives that lower the costs of development and "Pull" incentives, which provide the expected revenues. Our research focused primarily on incentives that have not yet been applied to the biodefense industry. We encountered a variety of opinions as to how applicable and successful various incentives would be at promoting MCM development. Key findings from interviews and literature include the follwing:

- No one push or pull incentive is sufficient to attract experienced companies to the MCME.
- There is no single best combination of incentives—the right package depends on context (policy and strategy decisions, requirements, technologies, pipeline maturity, etc.).
- Pull strategies should focus on increasing return on investment (ROI) through sustainable markets while push strategies should focus on cultivating partnerships and collaborations.
- Minimizing disincentives (e.g., lack of sustained and sufficient funding, government contracting process, lack of regulatory clarity)

may be enough to "tip the scale" and would "send a signal that the MCME is committed to collaborating with industry," potentially attracting additional private investors to MCM development.

Conclusion

Under the current policy of focusing on post-event response based on product stockpiling, opportunities to increase industry participation exist across the latter three segments of the MCME Business Model Framework. Alternative policies for product use, in turn, could have a cascade of alternative approaches. The most frequently cited opportunities for increasing the level and mix of involvement by pharmaceutical and biotechnology companies in the MCME include the following:

- Product Strategy: The USG should increase the emphasis on promoting multiple use products, platforms, and technologies with commercial applications.
- Players and Roles: Role assignments should focus on performer strengths, with innovator companies driving products through proof of concept (POC), and then partnering with experienced companies for late-stage development and manufacturing. USG should explore opportunities to promote collaborations, whether these are bilateral partnerships between companies or public–private partnerships.
- Incentives: The most critical incentive the USG can provide is to create a reliable market for MCM products. Additional incentives for consideration include priority review vouchers, new types of tax incentives for research and development (R&D) costs, and the funding of capital assets (equipment, manufacturing facilities, etc.) that can be leveraged for commercial purposes.

INTRODUCTION

Medical Countermeasure Development Challenges

The development of critical countermeasures is an ongoing challenge for the MCME. Most of the threats featured in the 2007 Public Health Emergency Medical Countermeasures Enterprise (PHEMCE) Implementation Plan for CBRN threats (Table E-1 below) require development programs to achieve either approval/licensure for new MCM products or new indications for existing products.[1]

Table E-1 Material Threat Determinations (MTDs) and Population Threat Assessments (PTAs) Issued to Date by the Department of Homeland Security

DHS: Material Threat Determinations (MTDs) and Population Threat Assessments (PTAs)	
Bacilus anthracis (Anthrax)	Marburg virus (Hemorrhagic Fever)
Botulinum toxins (Botulism)	Multi-drug resistant Bacillus anthracis (MDR Anthrax)
Burkholderia mallei (Glanders)	Radiological/Nuclear agents
Burkholderia pseudomallei (Meliodosis)	Rickettsia prowazekii (Typhus)
Ebola virus (Hemorrhagic Fever)	Variola virus (Smallpox)
Franciscella tularensis (Tularemia)	Volatile nerve agents [PTA only]
Junin virus (Hemorrhagic Fever)	Yersinia pestis (Plague)

The MCME Environment Is Complex, Challenging, and Uncertain

Unfortunately, the MCM landscape is plagued by uncertainty in an already complex and challenging field where the development of pharmaceuticals and vaccines is inherently risky, lengthy, and costly. Development cycles can take as long as 10 to 15 years and are conservatively estimated to cost $1.2 billion for small molecules and

[1] 2007 PHEMCE Implementation Plan.

$1.3 billion for biologics.[2] In contrast, Project BioShield's 2004 funding is only $5.6 billion dollars over 10 years of procurement for at least 13 MTDs and 1 PTA.[3] Additionally, most product candidates that enter clinical trials do not succeed, with only 13% gaining regulatory approval.[4] Even if these products reach the market, 70% fail to recoup their R&D investments.[5] Moreover, under the current business and incentive model has largely constrained the Enterprise to working with smaller, less experienced companies with little or no history of successfully developing, licensing, and producing products.[6] Compounding these challenges is the overarching uncertainty that touches virtually every aspect of the MCM value chain, including product requirements, market size, and regulatory constraints. The "market" for CBRN MCMs has largely been determined by intermittent stockpile procurements by the U.S. government under Project BioShield. This approach makes it difficult to anticipate government procurements and thus creates market uncertainties for companies. As with any other industry, high levels of uncertainty and low expected returns lead to decreased investor interest and pose major challenges to the MCME.

MCM development also faces regulatory uncertainty due to the Animal Efficacy Rule.[7] Under this guidance, efficacy is established through animal models rather than human populations for ethical reasons. To date, no novel products and only a two new indications of previously licensed products have been approved under the Animal Efficacy Rule.[8] Moreover, many threats do not yet have proven animal models available.[9] When models exist, they only provide a rough approximation

---

[2] DiMasi and Grabowski, 2007. Note that the biologics in the data set include monoclonal antibodies and therapeutic recombinant proteins, and do not include vaccines.

[3] Parker, 2007.

[4] Dimasi et al., 2009.

[5] Milken Institute, 2006.

[6] Matheny et al., 2007.

[7] In May 2002, the FDA published "Approval of Biological Products when Human Efficacy Studies are not Ethical or Feasible" [21 CFR 601 Subpart H, as well as 21 CFR 314 Subpart I for New Drugs]. This rule is more commonly known as the "Animal Efficacy Rule" or the "Animal Rule."

[8] Tucker, 2009. Pyridostigmine bromide was approved in 2004 as a pretreatment against soman, a chemical nerve agent. Hydroxycobalamin was approved under the Animal Efficacy Rule as a drug used to treat smoke inhalation as a countermeasure against cyanide.

[9] Gronvall et al., 2007.

of the efficacy of the treatment in human populations, as countermeasures developed using the Animal Efficacy Rule will remain untested in humans until used during an emergency.[10] Based on stakeholder feedback, the Animal Efficacy Rule pathway is seen as uncertain and riddled with risk by both large and small MCM developers, despite recent efforts by the FDA to provide additional guidance on the matter.[11] Furthermore, several interviewees questioned whether the FDA has enough focused resources and funding to manage MCM reviews vis-à-vis other products. SMEs also questioned whether reviewers in the Center for Biologics Evaluation and Research (CBER) and the Center for Drug Evaluation and Research (CDER) have the necessary public health and national security perspective to understand the unique requirements and context of MCM products required to evaluate the trade-offs and exceptions that come into play with their development.

Engaging Experienced Industry Players Has Proven Challenging

The mission inherently requires close collaboration between the USG and private sector companies. While stakeholder opinions varied about the type of industry participation needed, the majority believed that *some level* of experienced pharmaceutical engagement, particularly in late stage development, was a necessity to building a successful MCM development enterprise. Because of their size, smaller companies simply do not have the breadth of skills in development chemistry, process chemistry, manufacturing, etc., that can be found in larger firms, and this is a critical gap. As one subject matter expert (SME) explained, "When you're developing a new chemical entity and a new manufacturing platform that the Food and Drug Administration [FDA] is not familiar with, you need that experience. It's not that the smaller guys aren't smart enough, it's just that it's a game of breadth, not depth."

Despite the need for seasoned industry expertise, the MCME has had difficulty attracting significant interest from mid- and large-size pharmaceutical firms,[12] whose main barriers to entry are the high opportunity costs for potential time and money spent on CBRN MCM development activities in an industry where, according to one industry expert, "opportunity cost is everything." Unfortunately, the market

---

[10] Matheny et al., 2007.

[11] Guidance for Industry: Animal Models—Essential Elements to Address Efficacy Under the Animal Rule, 2009.

[12] Matheny et al., 2007.

opportunity that industry perceives for medical countermeasures is quite small, particularly for CBRN countermeasures, compared with the scale of other market segments that they address. As such, most MCM developers to date have been smaller biotech companies with fewer alternatives for development programs. As one SME explained, "While all potential partners consider funding as an incentive for participation, anticipated MCM funding is more likely to attract smaller biotech companies and academic labs." This is largely because the opportunity costs of small biotechs for undertaking MCM development activities are much lower than that of large pharmaceutical companies.

Inconsistent funding from the USG poses additional uncertainty and risk. More specifically, in FY09, $275 million was transferred out of the SRF for advanced research and development and $137 million was transferred out for pandemic influenza preparedness. In FY10, $305 million was transferred out of the BioShield Strategic Reserve Fund (SRF) for advanced research and development while $304 million was transferred out to the National Institute of Allergy and Infectious Diseases (NIAID).[13] Current FY11 requests would transfer $476 from SRF for advanced development. Transferring funds away from the SRF leaves less pull funding available for MCM acquisition and, more importantly, sends a negative signal to current and potential industry partners regarding the government's commitment to the MCM development mission. Congress has not articulated plans to reauthorize the SRF, creating unsettling market uncertainty for companies whose research and development programs are not expected to reach maturity until after the SRF is due to expire in FY2013.[14] Ongoing uncertainties about the level of annual appropriations make it difficult for MCME agencies to effectively manage multiyear MCM research and development programs and engage industry partners in the mission. Moreover, program managers are typically unable to fully fund all projects and build pipelines of concurrent candidates.[15] Because no one entity is necessarily responsible for funding end-to-end MCM development, it is important to ensure funding dollars are appropriately distributed across all stages of the product development pipeline.

Despite these challenges, most subject matter experts still believe that seasoned industry firms can play a role in MCM development,

---

[13] Optimizing Industrial Involvement with Medical Countermeasure Development: A Report of the National Biodefense Science Board, 2010.
[14] Matheny et al., 2007.
[15] Tucker, 2009.

though one representative from a leading pharmaceutical company representative noted, "If you were to search through our corporate strategy, the words 'biodefense' and 'medical countermeasure' will never appear." Stakeholders agreed that successful MCM development depends on establishing a breadth of capabilities not typically found in most small companies. A refined MCME business model may make the CBRN MCM opportunity more appealing to experienced pharmaceutical companies.

CURRENT UNITED STATES GOVERNMENT APPROACH TO MCM DEVELOPMENT

The Enterprise's CBRN program investments to date have primarily focused on biological threats. The approach that has been followed to date for developing MCMs against these threats can be summarized as follows:

- Policy decisions on how MCM products against biological threats will be used emphasize post-event response with stockpiled MCM products. These policies are described in the PHEMCE Strategy and Implementation Plan of 2007.[16] Such policies reinforce a perception that a successful developer would achieve only fixed, small volumes of MCM sales.
- The MCME is largely seeking to develop new MCM products (vaccines and therapeutics) for a diverse set of requirements (including special needs populations) against 14 agents for which MTDs and PTAs have been issued, most of which are for biologic threats. [17] Coordinating funding and development for these development projects across multiple USG agencies is a complex task. Recent progress has been made toward gaining cross-agency organization in an effort to help define and manage a single "Integrated Portfolio" for USG Biodefense MCM development.[18] This approach is intended to coordinate the biodefense MCM pipelines currently managed by the Biomedical Advanced Research and Development Authority (BARDA), NIAID, Department of

---

[16] PHEMCE Strategy and Implementation Plan, 2007.
[17] Ibid.
[18] Optimizing Industrial Involvement with Medical Countermeasure Development: A Report of the National Biodefense Science Board, 2010.

Defense (DoD)/Chemical and Biological Defense Programs (CBDP) and Defense Advanced Research Projects Agency (DARPA) and in turn achieve a more balanced pipeline of products from research and development through advanced development to FDA approval or licensure.

- The MCME is partnering with MCM developers to manage the entire MCM development chain from research, to development, and finally to production by employing a variety of incentives, including "push" mechanisms and "pull" mechanisms (e.g., grants, contracts, government/industry collaborations, liability protections, tax credits) and "pull" incentives (e.g., regulatory and exclusivity rewards, procurement contracts) that mitigate MCM developers' risk.[19]
- Historically, MCM advanced development by HHS depended primarily upon pull based incentives through Project BioShield. Under the Pandemic and All-Hazards Preparedness Act of 2007 (PAHPA), however, incentives that the U.S. government employs to achieve this mission now include a broader combination of "push" incentives (e.g., advanced development funding) and "pull" incentives (e.g., BioShield awards) that mitigate MCM developers' risk[20] in an attempt to help MCM development cross the perceived "valley of death"[21] in late-stage development. To promote advanced development and innovation, BARDA can award contracts, grants, cooperative agreements, and utilize other transaction authorities (OTAs). BARDA is also responsible for pulling MCMs through late-stage advanced development and into production by managing the Project BioShield SRF.

INDUSTRY NEEDS FOR MCME ENGAGEMENT

Figure E-1 illustrates several related factors that shape the level and mix of industry participation in MCM development. Policy decisions for how MCM products will be used help to set a concept of operations (CONOPS) and establish product requirements, which drive product strategies, development roles, and incentives needed. Plans for how MCMs will be supplied, distributed, and administered downstream have

[19] Matheny et al., 2007.
[20] PHEMCE Strategy and Implementation Plan, 2007.
[21] The "valley of death" is commonly used to refer to the costly late stages of development, requiring a commitment of significant resources for successful execution.

significant implications for upstream development activities. These considerations may expand or contract companies' assessments of MCM development opportunities.

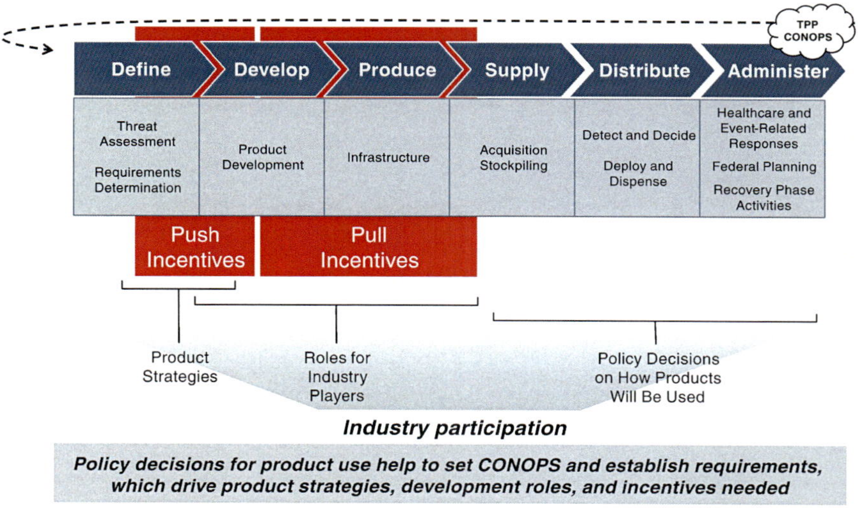

FIGURE E-1 High-level overview of how policies for product use, product strategies, roles for industry players, and push and pull incentives drive industry participation throughout the MCM Development Value Chain.

Given the factors that shape industry participation in the MCME, further consideration must be given to companies' "must-have" conditions prior to pursuing MCM development projects. A recent draft report by the National Biodefense Science Board (NBSB) Markets & Sustainability Working Group examines issues constraining or enabling industry involvement and highlights a series of recommendations for optimizing industrial support of medical countermeasure development.[22] Drawing from that NBSB report, an extensive review of current literature, and the stakeholder interviews, we have identified the three principal conditions that must be met in order to attract experienced pharmaceutical companies and VC funding to the MCME:

---

[22] Optimizing Industrial Involvement with Medical Countermeasure Development: A Report of the National Biodefense Science Board, 2010.

- **Requirements:** Simply stated, companies need to understand the requirements for the MCM the USG wants them to develop. Costs and risks are present at every step of the value chain, from quality control issues during R&D and manufacturing, to demand forecasting challenges, to competition from subsequent market entrants.[23] These requirements should help mitigate these expenses and uncertainties by including clear clinical research requirements, expected volumes specifications, detailed price information, and intended usage scenarios,[24] which drive the target product profiles (TPPs) for the countermeasures. With these requirements clearly stated, companies can then balance product requirements against their current capabilities and constraints in order to determine if an opportunity is worth pursuing.
- **Regulatory Clarity:** In addition to a clear set of product requirements, MCM developers must have a clearly defined regulatory path that can be navigated to success. Companies want to know what they have to do to achieve FDA licensure or approval. As one company representative explained, "The regulatory process is the most uncertain thing there is in biodefense and creates too much risk for pharma." The FDA provided draft guidance on this topic in January 2009 through its publication titled "Draft Guidance: Animal Models—Essential Elements to Address Efficacy Under the Animal Rule,"[25] though the exact requirements and restrictions of the regulatory path are still a topic of much debate. Some SMEs stated that the Guidance document should be repealed because it is overly restrictive. Comments have been received and are under review as the FDA works to finalize the document.

  It is important to note, however, that despite the importance of bringing these products to licensure, interviewees expressed the importance of not trying to "short-cut" regulatory standards for MCMs. They noted that it makes little difference how quickly a product gets to market if the public doesn't have confidence in its safety and refuses to accept it. Instead, interviewees voiced strong opinions that MCM developers should engage with the FDA early in

---

[23] GAVI Report: How can public-private partnerships accelerate the availability of vaccines for the developing world? July 2001.

[24] Hatchett, 2009.

[25] Draft Guidance: Animal Models—Essential Elements to Address Efficacy Under the Animal Rule, January 2009.

the process to validate plans and obtain guidance throughout the development process and submission preparation.

- **Return on Investments:** A final key consideration for any MCM developer is the need to offset the opportunity cost of participating in MCM development as opposed to developing other commercial products. As public companies with a fiduciary duty to shareholders, experienced pharmaceutical manufacturers in particular must ensure the returns are worth their investments of R&D dollars and time. One SME noted that a critical difference between pharmaceutical companies and other industries that regularly contract with the USG is the fact that "Wall Street expects a much higher rate of return from pharma companies."

Two key factors influencing the expected return of an MCM development project are margin and volume. Figure E-2 shows the financial attractiveness of an opportunity as a function of the expected margin and market size of the product. Today's MCME model anchors most development opportunities in the lower left quadrant, with small, periodic purchase volumes and low margins. Alternatively, competing projects tend to fall into the more fiscally attractive quadrants where they compete with MCMs on volume, margin, or both.

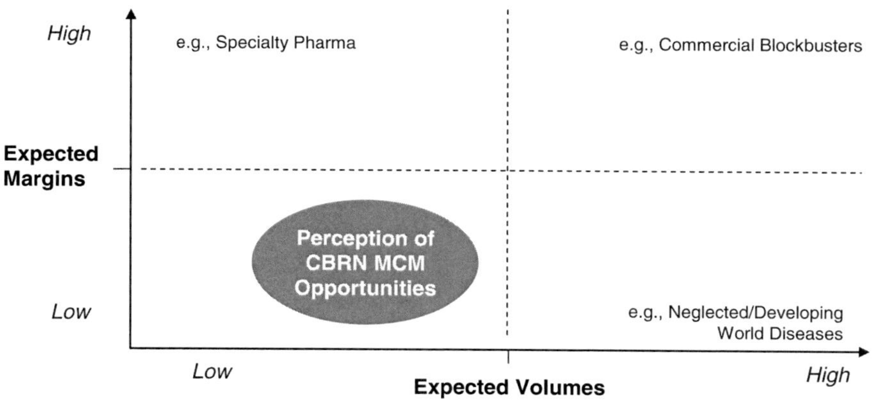

FIGURE E-2 CBRN MCM opportunities are perceived by industry as a low-margin, low-volume, and low-return investment.

Ideally, companies would like to be able to anticipate an expected return that is competitive against other potential projects before making an investment decision. At a minimum, however, they require some concrete indication that MCM development is considered important and credible assurance of a long-term commitment with financial backing, policies and incentives that support a predictable market. While MCM development projects will never be risk-free investments, industry players need to be able to anticipate an attractive return, to enable an informed decision relative to other opportunities.

According to stakeholder interviews and current literature, the three critical conditions outlined above are not being met. As such, many feel it is "almost impossible" for industry to justify pursing this mission as it does not present a sound business opportunity. As one SME noted, "You have an uncertain regulatory path to approval, the government determining procurement volumes, and the government reserving the right to change its mind. That makes it all kind of scary."

A MEDICAL COUNTERMEASURES BUSINESS MODEL FRAMEWORK FOR INDUSTRY PARTICIPATION

Resolving the observable disconnect between industry needs and today's MCME requirements will take more than tactical additions and adjustments to the current business model. The challenge at hand will not be solved with a short term solution, but rather through a series of policies and strategies coupled with tactical practices and incentives that will enable the current business model to evolve into a system that can achieve the desired state of operations.

An effective, mission-focused MCM business model depends on top-level strategic guidance, comprehensive operational planning, and successful tactical execution. The MCM business model framework spans all three:

- Strategic Planning: "What are we trying to achieve?"
- Operational Planning: "How will we achieve the mission, and with which products and organizations?"
- Tactical Planning: "What do we need to make it work?"

Framework Overview

The current approach to MCM development can be summarized as a business model composed of four basic components, each of which plays a pivotal role in defining, organizing, and executing MCM development (Figure E-3).

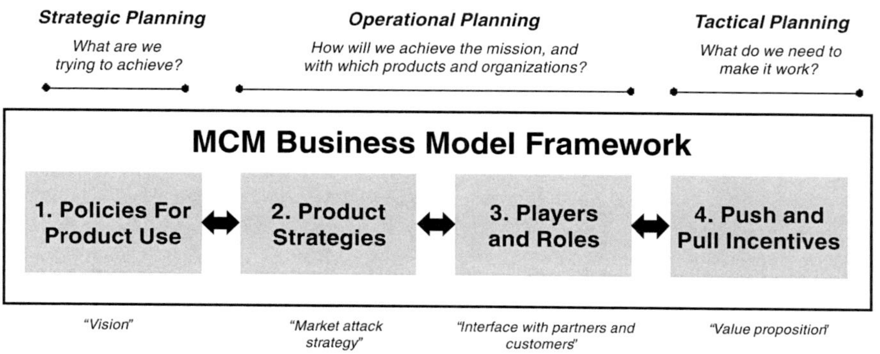

FIGURE E-3 MCM business model framework spans across strategic, operational, and tactical planning.

- **Policies for Product Use** (or "Vision"): Policy decisions for how MCMs are used (e.g., stockpiling, vaccination/prophylaxis, or forward deployment) that drive MCM product requirements (what, and how much)

- **Product Strategies** (or "Market Attack Strategy"): Development of new MCM products vs. new indications for commercial products, or indications for special populations

- **Players and Roles** (or "Interface with partners and customers"—[i.e., industry]): Company and customer types, roles in development, and structure of partnerships

- **Push and Pull Incentives** (or "Value creation"): Incentives provided by the government to increase industry interest in MCM development opportunities

Summary of Select Alternative Medical Development Models

This framework can be used to summarize components of alternative medical development models, including those for pandemic influenza, neglected diseases, and radiological and nuclear threats, as well as biological threats. These examples serve to highlight how policy decisions and product strategies have implications for industry roles and incentives needed. Table E-2 provides an overview of these models based on the framework.

TABLE E-2 MCM Business Model Applied to Pandemic Influenza, Neglected Diseases, Rad/Nuc Threats, and Biological Threats

Examples	1. Policies for Product Use	2. Product Strategies	3. Players and Roles	4. Push and Pull Incentives
Pandemic Influenza	Broad vaccination campaign for H1N1 vaccine	Extension of commercial vaccine for seasonal influenza	Performers: Experienced vaccine manufacturers responsible for development, production, and licensure of their own product. Customers: USG, State and Local governments	Significant pull incentives (procurement contracts) for H1N1 Building on Experience from development contracts since 2006
Neglected Diseases				
Radiological/ Nuclear Threats (Rad/Nuc)	Vendor or customer managed where feasible; Stockpiling where necessary	New indications for commercial products (primary approach) New products for biodefense only	Performers: Government conducting studies for Rad/Nuc indications for companies to file Customers: USG	Simplified contract management See section 9.1: "Government develops MCM/Company owns IP"
Biological Threats	Stock-piling MCM's for	New MCMs for	Performers: companies	Push: NIAID grants and

	post-event response	individual biological threats (primary approach to date)	responsible for research as well as development and production of their products; these have primarily been innovator companies Customers: USG	contracts; advanced development funding Pull: BioShield SRF procurement contracts

POLICIES FOR PRODUCT USE

The first component of the MCM Business Model Framework is a set of policy decisions that drive the rest of the business model. As described in the summary of the current approach for biological threats, the policy emphasis to date has been on post-event response using stockpiled MCM products.

Alternative policy decisions would have significant implications on the usage scenarios and CONOPS for MCM products. These would likely translate into changes in the volume of MCMs required (and thereby a potentially larger addressable market) and in the target product profile of products required. For example, if the decision were made to follow a policy of pre-event prophylaxis against a certain threat, companies could expect that sales of relevant vaccines could be greater than if the products are stockpiled.[26] Similarly, a policy of forward deployment of MCMs stockpiled in-home[27] could increase expected volumes. These policies could also expand potential buyers beyond

---

[26] Actual demand for pre-event prophylaxis is dependent on the public's willingness to use the countermeasure. SMEs noted that the public's receptivity to a new MCM product will depend on the perceived safety of the product, the public's trust in the government and MCM developers, and on the perceived expected probability and impact of a biothreat incident.

[27] Global Security Newswire, Former HHS Official Backs Home Kits for Biodefense, 2010.

agencies such as BARDA and the CDC to also include state and local governments, the DoD, and even individual citizens.

PRODUCT STRATEGIES

The second component of the MCM Business Model Framework provides a means for segmenting MCM products based upon development approach strategies and target populations. MCM products can be organized into four categories:

1. New Products with No/Limited Commercial Markets
2. New Products, Platforms, and Technologies for Flexible Defenses
3. New Indications for Existing Products
4. New Indications for At-Risk Populations

While each category presents a unique set of advantages and challenges, each one also plays an important role in maintaining a robust portfolio of MCM products and represents a different scope and scale of industry involvement.

New Products with No/Limited Commercial Markets

In the absence of related products that may be leveraged, CBRN MCMs must be developed and manufactured essentially for a single customer, the U.S. government, with potential for limited purchases by foreign governments. Under these circumstances, development must begin with very early stage research and progress through all of the subsequent MCM development stages. The associated time, cost, and risk introduced under these circumstances is a large disincentive to potential investors and industry collaborators. Most of the current biological MCMs fall into this category. These countermeasures address biothreats with unique biological pathways and consequences.

New Products, Platforms, and Technologies for Flexible Defenses

The MCME is interested in acquiring broad-spectrum products, platforms, and technologies that can have both commercial and biodefense applications, enabling biotechnology firms to rely more on commercial opportunities to generate adequate ROI as opposed to depending exclusively on government procurement.[28] Additionally, as biotechnologies become more powerful and accessible, there will be an increased range and severity of potential threats, whose growth may outpace the growth of "one bug, one drug" countermeasures that can be stockpiled.[29] The White House's Homeland Security Presidential Directive 18 and the PHEMCE Strategy and Implementation Plan of 2007 both recommended addressing this problem by developing broad spectrum countermeasures and platform technologies.[30] One SME touted the value of broad spectrum products, platforms, and technologies, stating, "I want to have multiple plays because I know most products fail." Despite widespread enthusiasm for this product strategy, broad-spectrum approaches have so far had limited success beyond new indications for some antibiotics.

New Indications for Existing Products

In some cases, it is possible to build upon the research and development of commercially available products to develop MCMs. One interviewer remarked that this is generally feasible when a potential biothreat causes symptoms or other biological reactions similar to those of diseases for which commercial drugs are already available. In such cases, it may be possible to leverage the discoveries, technologies, and other IP associated with a commercial drug to jump start the development of a new MCM. In theory, only late stage development work would be required to license the commercial drug as an MCM by showing efficacy under the animal rule, since safety will have already been proven, although additional clinical studies may be required for new dosing schedules. Additionally, the ongoing commercial market helps ensure a steady demand for the product, thereby alleviating some of the risk associated with inconsistent USG procurement and providing a more attractive case for investment by private companies. Thus this strategy carries potentially significant cost and time advantages, assuming a suitable commercial product is available.

Examples of products in this segment include some MCMs developed to address radiological and nuclear (Rad/Nuc) threats.

---

[28] PHEMCE Implementation Plan, 2007 and White House's Homeland Security Presidential Directive 18, 2007.

[29] Matheny et al., 2007.

[30] PHEMCE Strategy and Implementation Plan of 2007 and White House Homeland Security Presidential Directive 18.

Because of the strong similarities between cancer activity and Rad/Nuc exposure, the oncology drug market has proven to be a powerful source of potential Rad/Nuc MCMs. Rad/Nuc MCMs have been developed by modifying commercially available oncology treatments to meet requirements for Rad/Nuc countermeasures.

New Indications for At-Risk Populations

In order to ensure the MCM portfolio provides comprehensive coverage for all individuals, a fourth segment of MCMs encompasses those products specifically tailored for "at-risk" populations among the general public, be they new indications for available products or newly developed products. According to section 2802 of the Pandemic and All-Hazards Preparedness Act (PAHPA), at-risk populations are defined as "children, pregnant women, senior citizens and other individuals who have special needs in the event of a public health emergency, as determined by the Secretary."[31] Once an MCM is licensed with a label that meets an intended CONOPS, it may need to be augmented to suit the needs and requirements of these at-risk populations. SMEs suggested that there are few development strategies[32] that are unique to MCMs for special populations and that they should simply be considered as development programs for additional indications. Incentives such as those currently in place for pediatric indications for commercial products could be pursued.

PLAYERS AND ROLES

The third component of the MCM Business Model Framework addresses operational and organizational roles for successful MCM development and procurement. The MCME encompasses a diverse set of stakeholders, including pharmaceutical and biotechnology companies, federal agencies, state, local, and tribal governments, public health organizations, and, importantly, individual citizens. Collectively, these stakeholders have a broad range of roles including defining product requirements, conducting R&D, manufacturing, procurement, and distribution. In addition to the current approach described below, there

---

[31] Pandemic and All-Hazards Preparedness Act, Public Law 109–417, 120, Stat. 2831 (2006).

[32] Vanchieri et al., 2008.

are a number of alternative ways these players could work together to achieve the mission of bringing new MCMs to licensure. Several models for this MCM developer base are explored in section 9.1. Similarly, the MCM *customer* base can also vary depending on the policies and product strategies in place. Potential implications for MCM customers are discussed in section 9.2. Throughout the various models, the role of the USG in the developer base is generally one of either providing incentives directly to partners, or catalyzing interactions among industrial and academic partners. Regardless of its role on the performer side, the USG is a primary customer in each case. As expected, each alternative presents its own set of benefits and weaknesses.

MCM Developer Roles

- ***Research and Development by Small Innovator Companies: Current Predominant Approach***
  As discussed in previous sections, the MCME's existing incentive structure and current market challenges have primarily resulted in the USG being successful in enticing small, private sector entities to develop MCMs through licensure and produce them for USG procurement for stockpiling, distribution and potential use. Unfortunately, these small biotech ventures may lack the experience, expertise, and other general resources required to successfully complete the mission within time, cost, and quality goals, as noted in interviews with SMEs. This is particularly evident with respect to advanced development and production.

- ***Research and Development by Experienced Pharmaceutical Companies***
  Under this model, experienced pharmaceutical companies take responsibility for the end-to-end discovery, development, scale-up and manufacturing of new MCMs. This approach could entail adapting current commercial products to fit the requirements of a bioterrorism threat or developing novel products. The USG plays a relatively hands-off role, as the bulk of the project management and execution responsibility rests with the pharmaceutical company, as they are experienced at managing the entire length of the development value chain. The strength of this approach is that it takes advantage of the most knowledgeable experience base available and maintains consistency throughout all phases of

development. The dominant weakness is that the opportunity cost to the industry player is tremendous and likely requires significant levels of USG incentives. Moreover, experienced pharmaceutical companies are not necessarily in the best position to execute early discovery work, as small companies tend to drive innovation.[33] As one SME noted, "*discovering* good drug candidates is not a prerogative of big pharma."

Successful examples of experienced pharmaceutical industry participation in the CBRN space do exist, as evidenced by the Pandemic Flu vaccine program. Here, experienced industry performs virtually all phases of production, and supports the annual development of new seasonal flu vaccines. It must be noted, however, that seasonal flu represents a recurring revenue stream for these industry players, as opposed to a one time production for stockpiling, highlighting the opportunity for incorporating flexible defenses in engagements with industry manufacturers.

- ***Government Develops MCM/Company Owns IP***
  Under this model, the USG assumes responsibility for late-stage[34] development of MCM products, building on the early stage work of biotech companies. The USG plays a very active role in this model, overseeing all late stage development activities using USG resources and facilities. Strengths of this model include the fact that the government does not have to invest heavily in the high-risk, early stage research and discovery efforts, as USG involvement does not commence until products are ready for late-stage development. Additionally, the USG is able to maintain a high degree of control over the critical late stage development and regulatory activities. The model suffers from companies' perception of risks involved in putting a successful commercial candidate back through investigational studies for an alternate indication, given the potential for adverse effects or other negative outcomes. Thus, when applied to a commercial product already on the market, some companies may be hesitant to engage when the product at hand has (or potentially could have) a lucrative commercial market application.

---

[33] Munos, 2009.

[34] Late stage development typically represents Phase II clinical trials and beyond.

A recent example of this approach can be found in the Rad/Nuc MCM development program. Here, the USG has demonstrated it is possible to develop a pipeline of Rad/Nuc MCM candidates on a very limited budget by drawing on successful products and candidates from private companies and academic researchers. While the Rad/Nuc program does not fund programs directly, it provides in-kind services to companies who are interested in furthering their product's scope by adding biodefense indications.[35] One notable factor of these collaborations is that they do not typically require government contracts to reach licensure and are therefore not subject to Federal Acquisition Regulation (FAR), making it simpler, and "less painful," for private companies to engage with the USG.

- ***Government Owns IP/Company Develops***
  Another option for government and private sector collaboration is to outsource government owned MCM compounds for development by private companies, while the USG retains ownership of the IP. This model allows the government to exercise control over valuable, and potentially sensitive, intellectual property, while taking advantage of a dedicated, specialized team of scientists. A notable example of this model is seen in the DynPort Vaccine Company (DVC) and its support for the DoD's Joint Vaccine Acquisition Program (JVAP). For example, under this contract, the U.S. Army Medical Research Institute of Infectious Disease (USAMRIID) was involved in identifying the suitable protein antigens to create the rF1V plague vaccine, and subsequent efficacy testing, while DVC manages the vaccine's advanced development up to and including possible licensure by the FDA.[36]

- ***Experienced Company/Innovator Company Partnership***
  The goal of this model is to take advantage of the unique and valuable contributions both experienced pharmaceutical and innovator companies can make toward MCM development by encouraging them to form product development partnerships. Under this strategy, experienced pharmaceutical companies support early stage development of smaller, more innovative biotechs by providing financial backing, resources, and expertise. Once the product begins

[35] Hatchett, 2009.

[36] CSC News Release: CSC's DynPort Vaccine Company to Continue Plague Vaccine Development, 2008.

to mature, the large pharmaceutical partner supports or manages late stage development and manufacturing. The role of the USG in this model is to facilitate and enable the formation of such partnerships. Many SMEs surveyed maintained that, despite big pharma's development expertise, they are not necessarily the best innovators, and "don't do the early stage stuff well." As such, many SMEs were supportive of this type of partnership model and saw it as a logical, valuable distribution of labor.

- ***Public Private Partnership (PPP)—Product Development Partnerships or Manufacturing***
  A PPP is any relationship where public organizations partner with private organizations to share personnel, intellectual property, facilities, equipment, technologies, and other resources. Often no one sector has the skills or resources to address a complex public health challenge single-handedly, and commercial incentives alone may be insufficient to trigger private investment. Partnering provides an opportunity not only to deploy the right skills and resources, but also to share the risks. Under this model, the public and private sectors contribute in different, yet complementary, ways to the partnership. A public sector partner typically articulates the need, defines the vision, and makes initial commitments to mobilize the partnership. This is achieved through sustained and targeted public sector investments in technology, human capital, public health systems, or infrastructure. Private sector partners often provide expertise in applied technology development, commercialization, systems integration, human capital, and the application of market-proven business practices and systems. Examples include product development partnerships focused on effective drug and vaccine development, such as the Medicines for Malaria Venture (MMV), and Access PPPs like the Global Alliance for Vaccines and Immunization (GAVI Alliance), aimed at improving access to medicines by targeting populations.

  PPPs offer a wide array of potential benefits:

  Offsetting Opportunity Cost—As public companies with a fiduciary duty to shareholders, pharmaceutical firms must structure their portfolio to pursue only the most profitable projects. However, government support of PPPs may allow pharmaceutical companies to support MCM projects critical to homeland security despite weak capital markets.

Capacity Management—This PPP approach eliminates the dependence on industry manufacturing capacity and allows pharmaceutical companies to support the MCME without interrupting or compromising commercial production in their own facilities (provided the PPP operations have dedicated facilities and resource teams). Furthermore, a PPP with flexible manufacturing capacity could provide the ability to produce a wide range of products in relatively low volumes, making it ideally suited to support the MCME.[37]

Technology and Talent Development—A PPP also provides the opportunity for pharmaceutical companies to explore and test new technologies that could benefit both biodefense and commercial projects. SMEs noted that major pharmaceutical companies are often hesitant to pursue a new technology that could improve a marketed product for fear of encountering a complication in the development, such as a clinically adverse event, that could affect the marketing of the commercial product. However, they would be more willing to explore and apply new technologies to the biodefense application (using government funding) and then use those technologies for commercial products once the technology is proven and the systems are in place. Thus the opportunity to explore new technologies for MCMs with government funding and then build upon those technologies in the commercial market could be very attractive to industry. Similarly, PPPs provide an incubator for analytical talent for the industry firm. Essentially, the PPP serves as an instrument for human capital growth—developing employees with new scientific and technical based skills that could be applied to the parent organization.

The majority of interviewees were supportive of creating PPPs to research, develop, and manufacture MCMs, and examples of such collaborations have shown positive results. Recent examples of PPPs at work include the Cystic Fibrosis Foundation, the Michael J. Fox Foundation for Parkinson's Research, and the Multiple Myeloma Research Foundation. A commonly noted disadvantage of PPPs is

---

[37] Fuerst et al., 2009.

that products and technologies resulting from these agreements may face complicated disputes over intellectual property.[38] Thus careful consideration and forethought must be given to ensure all parties agree on the IP ownership plan to ensure downstream IP discrepancies do not hinder a PPP's effectiveness.

- ***Outsourced Development: Virtual Pharmaceutical Company*** Another alternative for the MCM development performance base is built on the notion that multiple organizations can collaborate to the point that a "virtual pharmaceutical company" is created. The goal is to form a unified network of companies capable of end-to-end execution of all aspects of the research, development, licensing, and manufacturing of an MCM. More specifically, BARDA or another participating government agency can directly manage MCM development by linking smaller biodefense discovery companies with contract manufacturing organizations (CMOs), contract research organizations (CROs), specialized facilities that test biothreat formulations and other types of intellectual property or technical resources.[39] Several industry experts remarked that the virtual pharmaceutical company model requires a manager with experience at a large pharmaceutical company who understands all development phases, particularly for processes related to late-stage development. The virtual pharmaceutical company model allows specialization and outsourcing of each segment to best available skills at lower costs to attain efficiencies. These efficiencies are especially apparent when the research segment is outsourced to the party that can most successfully complete segment's goals.

The primary weakness of the virtual pharmaceutical company model applied to MCM development is that it is essentially constrained by the current capabilities of existing pharmaceutical, biotechnology, and support companies. In other words, this approach does little to stimulate the development of new technologies, infrastructures, and personnel not yet in existence unless additional funding is provided for such purposes.[40] Similar to the public-private partnership model, intellectual property ownership can be a point of contention.

---

[38] Matheny et al., 2007.
[39] Scannon, 2008.
[40] Ibid.

MCM Customer Roles

The customer base of the MCME is largely determined by the policies and product strategies in place. For example, the current approach of post-event response using stockpiled MCMs makes the USG (and to a lesser extent state and local governments) the sole customer for MCMs and generally limits purchases to one-time bulk acquisitions and warm base manufacturing contracts. As alternatives to the current MCM business model are considered, it is important to note that any changes to policies and product strategies will have direct implications to the size and characteristics of the MCME customer base. Depending on the policies set forth going forward, MCM customers could expand beyond the USG to include state and local governments, private citizens, international governments, and global health organizations.

While numerous scenarios are possible, pre-event prophylaxis, pre-deployment of MCMs, and international pooled procurement stand out as alternative approaches that could have significant impacts to the MCM customer base:

- Pre-event prophylaxis could provide a much larger, more predictable demand for vaccines.
- A pre-deployment strategy of distributing therapeutic MCMs as home supply kits would enable citizens to maintain personal stockpiles in their homes. Provided USG does not decide to purchase and distribute all MCMs, this approach would likely introduce private citizen as customers, creating an even larger customer base as each citizen would bear some responsibility to retain personal coverage rather than depending upon national reserve stockpiles.
- Finally, the MCME may expand the customer base for US-made biodefense drugs and vaccines to allied countries to aggregate demand for MCM products. The USG has participated in the Global Health Security Initiative (GHSI), composed of health administrators from the G-7 countries and Mexico to address CBRN threats and expand access to countermeasures.[41] If such a multinational pooled procurement collaboration were to occur,

---

[41] See www.ghsi.org/.

the customer base would be expanded even further to potentially include foreign governments and healthcare facilities.

PUSH & PULL INCENTIVES

The fourth component of the MCM Business Model Framework is a set of incentives for making the MCM development opportunity more appealing to experienced pharmaceutical companies. This white paper has explored multiple incentives that have been proposed or implemented in other contexts but not yet applied to the MCME. The research effort explored incentive models from a broad range of sectors, including those outside the life sciences industry. While other sectors offer relevant tools and concepts to inform incentive structure design in the MCME, no "off-the-shelf" models for MCMs were directly applicable. A summary of insights gleaned from other sectors is included in the Appendix. This section will focus on incentives deemed most relevant to the MCME mission, highlighting potential benefits and weaknesses relative to their applicability to MCM development.

The incentives presented below are grouped into "Push" incentives that lower the costs and risks of development and "Pull" incentives, which enhance the expected revenues. Most interviewees suggested that a combination of incentives could be effective in attracting companies to the MCM Enterprise and that these mechanisms could be used to augment the current MCM model. One SME summarized the issue as follows, "One thing is for sure in establishing such external incentives: one size does NOT fit all because of the varying sizes, capabilities, and capacities of companies who could address MCM development." Combinations of push and pull incentives may be sufficient, depending on policy and strategy decisions for each MCM program, as well as where the incentives are applied along the development chain.

"Push" Incentives

Push mechanisms are intended to incite interest, action, and investment into scientific research on a particular problem by lowering the cost of research & development. Push incentives generally consist of tools for providing funding or other resources to make MCM R&D less expensive,[42] including grants, tax credits, help developing clinical trial infrastructure, and assistance during the regulatory review process. Essentially, push incentives provide assistance to participating industry partners in order to lower development cost or risk.

Currently, the USG provides several push mechanisms to engage MCM developers (e.g., grants, contracts, government/industry collaborations, liability protections, and tax credits). For example, the Pandemic and All-Hazards Preparedness Act (PAHPA) allows HHS to make advanced payments worth up to 50% of the value of a BioShield procurement contract to MCM developers before the delivery date of the product, provided that all milestones are successfully completed. [43] PAHPA also provides "limited antitrust exemption" that enables companies to participate in the joint development of an MCM. [44] MCM candidates, if approved by the FDA, could also benefit from tax credits if granted Orphan Drug Designation (ODD), [45] which is reserved for approved drugs that have a U.S. market of less than 200,000. All MCM R&D costs are also eligible for a research and experimentation (R&E) tax credit of 20% on qualified expenses.

One general criticism of push mechanisms is that money is applied without a guaranteed outcome. As such, there may be little impetus to move research into a clinically approvable product.[46] This is less of a problem for an association like the Global Alliance for TB Drug Development, which actively manages all phases of the development process to ensure that research is focused on outputting an effective and safe vaccine. The most significant problem with push incentives, according to several of our SMEs, is that pharmaceutical companies want a market, not lowered development costs. While push mechanisms certainly help to incentivize industry participation, they may be inadequate by themselves, particularly for large pharmaceutical companies with higher opportunity costs.

---

[42] Matheny et al., 2007.

[43] Pandemic and All-Hazards Preparedness Act, Public Law 109-417, 120, § 406, Stat. 2831 (2006).

[44] Pandemic and All-Hazards Preparedness Act, Public Law 109-417, 120, § 405, Stat. 2831 (2006).

[45] Grabowski, 2005. The Orphan Drug Act of 1983 has helped bring more than 200 drugs and biological products for rare diseases to market. The rate of orphan drug approvals has increased tenfold since 1983.

[46] Brogan, D., and E. Mossialos, 2006.

"Pull" Incentives

Pull mechanisms seek to encourage private companies to develop MCMs by providing a reward if the desired goal is achieved. This can be done in several ways. Pull incentives can create a profitable market for the MCM, allow a regulatory or marketing reward to be applied to another more valuable product, or provide a capital asset that can be leveraged for a commercial product. Unlike push incentives, pull incentives generally pay out when the developer has reached a particular milestone, such as product approval or licensure, and in some cases, production of an FDA-approved/licensed product.

Currently, the MCME's main pull mechanism is Project BioShield, a $5.6 billion SRF aimed at creating a market for vaccines against bioterrorism agents.[47] Thus far, the funds have not succeeded in attracting large pharmaceutical companies to MCM development, but have instead engaged smaller developers with limited infrastructure or experience bringing a product to market.[48]

A common weakness in any pull mechanism requiring commitment of government funds is that it is very difficult for the government to confirm how much incentive a private company requires to begin an R&D program. Government subsidies can "crowd out" private capital, particularly for dual use products, that the company would otherwise use in the same R&D programs.[49] Researchers have postulated that policy makers should be aware that some amounts of tax dollars are replacing private funds without a corresponding net increase in R&D activity for MCMs.[50]

NEW PUSH AND PULL INCENTIVES

This white paper explored various push and pull incentives from other sectors that have not yet been used in the biodefense industry. The following sections detail potential benefits, weaknesses and implementation challenges in applying these incentives to the MCME.

---

[47] Gottron and Fisher, 2004.
[48] Matheny et al., 2007.
[49] Maurer, 2009.
[50] Ibid.

These sections will explore the following new push mechanisms:

- Enhanced Tax Credits
- Access to Intellectual Property
- Access to Technology, Capacity, and Regulatory Services

These sections will explore the following new pull mechanisms:

- Advanced Market Commitments (AMCs)
- Priority Review Vouchers (PRVs)
- Market Exclusivity Rewards and Patent Extensions
- Prizes
- Leverageable Capital Asset Investments

Push Incentives

Enhanced Tax Credits

Tax credits are a means of incentivizing an activity for which there is an insufficient reward or return on investment (ROI); they can also compensate the developer for creating products that serve the public good, which in this case is the development of MCMs. Tax credits can be applied in a variety of ways to incentivize MCM R&D in general, as well as encourage more activity surrounding certain aspects of the MCM development process. Below are examples of some existing tax credits:

- *Research and Experimentation Tax Credits:* R&D activities are a public good that have broad social and economic gains.[51] Most companies now receive an R&E tax credit of around 20% for qualified R&D expenses to incentivize R&D activity.
- *Orphan Drug Tax Credit:* Currently, vaccines and therapeutics developed to treat rare diseases can get a 50% tax credit on clinical trial costs if granted Orphan Drug Designation (ODD) by the FDA.[52] Most MCMs are eligible for ODD in the United States because of their low disease prevalence. According to Dr.

---

[51] Doremus, 2008.

[52] Orphan Drug Act of 1983 P.L. 97-414.

Marlene Haffner at the 2010 IOM Conference, four MCMs have been approved for ODD since the conception of the program.[53]

Existing R&E and Orphan Drug tax credits are likely insufficient to incentivize adequate engagement in MCM development. New tax credit designs beyond what is currently available can be considered as well:

- *Tradable tax credits:* For firms that do not have significant streams of current income (e.g., most biotech firms), tax credits will yield little or no returns. The above tax incentives could be made transferable to address this issue, and unprofitable biopharma firms could trade their credits to profitable firms. Alternatively, tax credits could also be made deferrable to a future time when the firm becomes profitable.[54]
- *Strategic Partnership Tax Credits:* The USG can issue a tax credit for experienced pharmaceutical companies who partner with innovator biotechs to develop MCMs. This may motivate large companies to actively seek out innovator companies who are working on promising technologies.
- *Manufacturing Facilities Tax Credits:* The USG can institute a new investment tax credit for the construction of new R&D and manufacturing facilities for MCM production in the United States. Tax credits for manufacturing facilities may be particularly useful for vaccine manufacturers, as manufacturers must take years to build and validate new manufacturing facilities before the vaccine can be approved, incurring significant capital expenditure and risk along the way.[55]

Potential Benefits

For USG: Tax credits can be used to incentivize certain industry behaviors that are beneficial for MCM development. There is evidence to suggest that tax credits are relatively efficient at incentivizing R&D activity. One study of science and technology econometrics found that one dollar in tax credits resulted in one dollar of investment in R&D.[56]

---

[53] 2010 IOM Conference Proceedings.
[54] Matheny et al., 2007.
[55] Berndt et al., 2008.
[56] Audretsch et al., 2002.

For Industry: Tax credits provide substantial savings to R&D expenses without the formality of contracting with the USG.

Potential Weaknesses

Creative accounting could allow only tangentially related R&D activities to receive tax credits. Careful restriction of tax credits to include only activities related to MCM development would help to curb improper accounting.[57]

Implementation Issues

Implementing new tax credits would require congressional action, and the US government would pay for tax credits through reduced tax revenues.

Access to Intellectual Property

The National Institutes of Health (NIH), the Centers for Disease Control and Prevention (CDC), U.S. Army Medical Research Institute for Infectious Diseases (USAMRIID), and similar agencies have a wealth of intellectual property (molecular libraries, animal models, methods, techniques, etc.) that can be of interest to researchers in the private sector. There are multiple ways to enable access to such IP in the MCM development space. One option is to form a public–private partnership where government agencies share available molecular entities for MCMs with their commercial counterpart. Another option is to maintain open source access to government patents and patent applications for certain MCMs. Much like an open source programming language, once a product becomes open source, any authorized researcher can work on the product without infringing upon patent rights. Participants frequently join open source ventures for nonmonetary incentives such as ideology, to gain reputation, or to advertise skills to potential employers. Projects like LINUX have been remarkably successful and have generated a significant amount of interest.[58] In the healthcare space, GlaxoSmithKline (GSK) recently deployed a similar model whereby

[57] Maurer, 2009.
[58] Linux website.

they granted access to 800 patents and patent applications for researchers working to understand neglected tropical diseases in Least Developed Countries.[59]

Potential Benefits

For USG: Setting up the infrastructure to disseminate government-owned IP is relatively inexpensive, and could generate interest and activity from experienced scientists. According to one SME, there is a big difference between the leading pharmaceutical companies and the scientists who work for those companies and, in his opinion, the current organization of the corporate pharmaceutical world has many scientists working on "boring" projects that "they don't believe in." By simply providing opportunities for experienced scientists to work on interesting problems, he believes one can harness a significant amount of valuable scientific manpower. "Innovative folks love to work on scientific enigmas. The thrill of cracking a scientific challenge provides a lot of motivation for them." As such, pursuing an open innovation approach is one way to jump start this type of widespread collaboration and generate a lot of activity in the early stage MCM development space. Open source innovation allows the USG to capitalize on the experienced resources that make large pharmaceutical companies so successful at late stage development. As one SME noted, "If you're not going to use big pharma, then you have to have big pharma people."

For Private Sector Participants: SMEs remarked that increasing private sector access to intellectual property (IP) can help commercial participants lower product development costs by accelerating the discovery process.

Potential Weaknesses

Open source access to IP may push research to a wider variety of scientists, such that quality standards may not be as easy to uphold.[60] Without a sponsor overseeing work, researchers may misappropriate funds, avoid work, or misstate their results. Additionally, open source methods work best for research requiring little capital or materials and a lot of labor, which may explain why Open Source methods are generally

---

[59] Witty, 2010.
[60] Munos and Chin, 2009.

associated with software.[61] Sharing IP may be effective in promoting early stage drug discovery, but does not incentivize industry players to undergo late stage development, which is generally expensive and time-consuming. Successful licensure of MCMs conceived through open source early stage research would likely require the support of a PPP or other commercial operation to see candidates through late stage development and into production.

Implementation Issues

Political Challenges: Open source approaches may not be favored if the USG prefers to hold certain biodefense-related IP confidential for security purposes. There are few other political challenges to creating a PPP or open source forum to share government IP, as implementation requires few tax dollars, and relies primarily on volunteers and corporate contributions.[62]

Incentivizing Open Source Participation: Scientists may be unwilling to partake in an enterprise for which they are afforded no credit. It would help if the existing legal framework could be supplemented or modified to support scientific micro-contributions, such as an online registry that would allow scientists to log and stamp their contributions. If the product becomes a commercial success, these inventors can receive a share of profits based upon their relative contribution.[63]

Access to Capacity, Technology, and Regulatory Services

The USG can directly or indirectly provide access to facilities and technical and regulatory services to facilitate the development of MCMs amongst various market players, including public, private, and not-for-profit companies. The most useful mechanism for delivering such access would be through the creation of a public private partnership wherein a dedicated, flexible operation exists to support the development of biodefense products by providing immediate access to development technology, manufacturing capacity, and experienced human capital.

---

[61] Maurer, 2005.
[62] Ibid.
[63] Munos and Chin, 2009.

Potential Benefits

For USG: The USG can benefit from the expertise of resources from across the spectrum of the drug development process. These individuals and companies will likely be more willing to engage in MCM development once capital investments, technical capabilities and other barriers to entry are removed.

For Industry: Pharmaceutical firms can benefit from capital investments, technology resources or help navigating the FDA approval process. One industry expert remarked that facilities provided through public private partnerships are also excellent incubators for new analytical talent for the company. New scientists can be placed on biodefense projects to access and learn new technologies and techniques from outside their own company.

Potential Weaknesses

The chief disadvantage of solutions and products developed using shared facilities and resources is that technologies resulting from these collaborations may face complicated disputes over intellectual property.[64] Additionally, our talks with industry experts indicate that the operation may not succeed if there is too much government oversight limiting the private partners' freedom to operate.

Implementation Issues

Soliciting Private Sector Needs: By offering access to capacity, technology, and services, the government is essentially bridging critical gaps in private sector capabilities. These gaps will differ on a case-by-case basis. As such, the participating government agency should perform exhaustive due diligence into what its private sector partners require, whether that is funding, personnel, technology resources, animal models, manufacturing capabilities, and/or regulatory insight.

Pull Incentives

---

[64] Matheny et al., 2007.

The pull incentives listed below are intended to enhance expected revenues of MCMs in development.

Advance Market Commitments

Private pharmaceutical companies are oftentimes reluctant to participate in the research and development of biologic and therapeutic products that target rare and neglected diseases because the market for these products is not large and/or affluent enough to be profitable. Advance market commitments (AMCs) are pools of funding used to guarantee a market price for these products as a means of "pulling" development along. More specifically, an AMC is a guarantee by governments or other sponsors to pay developers a minimum price per dose of a medical product purchased in the market up to a specified volume. Products meeting the specifications of the AMC and purchased in the commercial market are guaranteed a co-payment from the sponsor up to a specified volume of sales. Participating companies also make binding commitments to supply the drugs or vaccines at lower, sustainable prices after the depletion of government funds made available for the initial fixed price. In 2007, the Global Alliance for Vaccines and Immunization (GAVI), which includes the Bill & Melinda Gates Foundation, along with several donor countries, created the first AMC, valued at US$1.5 billion dollars, for a multiple pneumococcal vaccine. The AMC will provide 7 to 10 years of funding, after which a long-term price will be available to consumers at near marginal cost. As of October 2009, the GAVI AMC has succeeded in attracting four offers to supply the pneumococcal vaccine from four different suppliers.[65]

AMCs differ significantly from advanced purchase commitments (APCs) by guaranteeing a purchase *price* up to a specified volume of sales, as opposed to guaranteeing a purchase *volume*. Under the current BioShield program for example, the USG procures MCMs by setting a future *price and volume* for the product, whereas an AMC prespecifies a price per dose up to a specified volume, without making a specific volume commitment up front. AMCs also contract a lower, sustainable price for the product after the funding is depleted, while traditional procurement contracts have no such provision. AMCs are not directly

---

[65] Gavi Press Release, 2009 As of writing, UNICEF's Supply Division is evaluating the offers and an independent assessment committee is assessing vaccine candidates, which must meet the Pneumococcal AMC Target Product Profile (TPP).

applicable to federal stockpiling modality (e.g., USG only buyer), but could be very useful in pursing alternate CONOPS (e.g., pre-event programs, home stockpiling, state stockpiling) provided that individuals, families, or states are required to pay some portion for MCMs supported by USG co-payment.

Potential Benefits

For AMC Sponsors: If the AMC is successful, the sponsor of the AMC will be able to access the resulting MCMs in sufficient quantity, at an affordable price, and over the long term. Additionally, the sponsor would pay its part only for those doses actually sold in the market.

For MCM Developers: AMCs help create a predictable market where the actual MCM market is limited.

Potential Weaknesses

For AMC Sponsors: AMCs require the sponsor to create a viable market for the MCM. An AMC large enough to induce industry participation is uncertain but will likely be expensive. The GAVI AMC for a pneumococcal vaccine is $1.5 billion.

For MCM Developers: The long-term credibility of AMCs is a source of uncertainty and risk. Since R&D costs are extremely high, and if we assume R&D spans over 10 years, with another 10 years to recoup the initial investment, the perception is that this becomes a very long horizon. Pharmaceutical companies may be concerned with the government reneging on the promised offer, given that government priorities may change over the course of 20 years. This possibility is compounded by the significant likelihood that a subsequent entrant would create a clinically superior product that undermines the government commitment to the first generation product.

AMC participants also face high risk with respect to the demand for the MCM. Since an AMC does not commit volume for the MCM product, the MCM developer may face little or no demand under a pre-deployment or prophylaxis/vaccination scenario, which may undermine the viability of an AMC as a pull incentive. For example, under a pre-event scenario, if the general public is unwilling to take prophylaxis measures (i.e., an anthrax vaccine), the AMC may never get fully utilized.

Implementation Issues

Although AMCs have theoretical merit,[66] they face a unique set of implementation challenges. Sponsors should consider how to set the size of the AMC and the long-term price, manage subsequent entrants into the market when the initial entrant has already capitalized on an existing AMC, and navigate through any political challenges.

Setting the Size of the AMC: Setting the AMC size is difficult as the societal value of the product must be weighed against the program costs. Furthermore, companies and the USG must negotiate a price that will both incentivize the private sector to develop the product, as well as be fiscally acceptable for the agency sponsoring the AMC. If multiple products are purchased, the perception of risk is lower, and the resulting price offered to the developer may be lower as well.[67] To give an example of the magnitude of the cost, the necessary size of an AMC to incentive R+D from large pharmaceutical companies has been estimated to be around $3 billion for a malaria vaccine—enough to equal the expected revenue of developing one commercial drug.[68]

Long Term Price: The contractual price of the MCM after the initial AMC has run out does not necessarily have to be near marginal cost. Whereas a central tenet of using AMCs for the neglected disease market is to ensure affordability for patients in developing nations, patients in the developed countries arguably can afford a higher long-term contract price.[69] Establishing a fair, reasonable, and sustainable differential pricing scale for the AMC's long term tail price is a challenging exercise that remains a subject of much debate within the AMC community.

Subsequent Entrants: AMCs should be designed to allow for second- and third-to-market entrants, as it is unlikely that the first product to market will meet all the needs of targeted patients.[70] One example of a pricing scheme that may achieve this effect is a falling price structure, where the initial entrant is paid the highest price per dose, with prices falling over time for subsequent entrants.[71]

Market Factors: Sponsors must also understand the implications of the fact that AMCs remain untested for MCM development. The MCM market is fundamentally different from that of neglected diseases. Whereas the neglected disease market is one of high prevalence but low

---

[66] Berndt and Hurvitz, 2005.
[67] Towse and Kettler, 2005.
[68] Berndt et al., 2007.
[69] Plahte, 2005.
[70] Berndt and Hurvitz, 2005.
[71] Ibid.

expected margins, MCMs generally target treatments for diseases that have little or no market in the United States but potentially have significantly higher margins per dose. This fact will create complexities for how sponsors should think about potentially augmenting initial demand for the MCM, as well as contracting a long-term price after the depletion of the AMC.

Political Challenges: it is unclear whether or not the USG can implement AMCs under current authorities. Congressional action may be required to authorize the organization sponsoring the AMCs to commit public funding to an AMC.

Priority Review Vouchers (Regulatory Reward)

Priority review vouchers (PRVs) are transferable prizes that allow the holder to secure an FDA priority review for a product of their choosing. A priority review is essentially a commitment by the FDA to complete and act on the review of an application in 6 months, whereas standard reviews can take 10 months or longer. A PRV can be sold if its original owner does not wish to apply it to a product in its own pipeline. The holder of the PRV would be required to pay a user fee, which some have estimated to be up to $1 million dollars, to enable the FDA to add resources and personnel to expedite the review process. PRVs have been created for use in the neglected disease field where, under the Food and Drug Administration Amendments Act of 2007, companies who create MCMs for the treatment of 16 tropical diseases receive PRVs for the successful approval or licensure of their products.[72] Though eligibility is currently limited to tropical disease products, it is interesting to consider how the incentive could impact MCM development if biodefense products were also eligible.

Potential Benefits

For USG: PRVs generally have fairly low social costs and are more politically acceptable than patent extension vouchers. Whereas granting patent extensions would delay the introduction of generics to consumers, PRV vouchers allow consumers to benefit from having new drugs and vaccines earlier than under the regular review process.[73] Additionally, implementation costs for PRVs are nominal compared with larger-budget

---

[72] Grabowski et al., 2007.
[73] Ibid.

items such as AMCs or tax credits. SMEs estimate that it would cost about $1-2M for the FDA to go from a standard review to a priority review.

For PRV Recipient: PRVs potentially shorten the FDA review process by 6 to 12 months, getting the product to market sooner where it can benefit a greater number of patients.[74] PRVs are also transferable, and function as a saleable asset if the owner does not wish to use it on their own product. Besides extending the effective life of the patent,[75] drugs using PRVs to speed approval may have the "early-mover advantage"; this means that they can profit from consumers who are reluctant to use a newly introduced competitor if they are familiar with a current product.[76] If the PRV resulting from successful MCM development is applied to a blockbuster product, a PRV can potentially be quite valuable, as it can extend effective patent life significantly.[77] A PRV is estimated to be worth $300 million dollars when used on the top decile of compounds currently on the market, and $100 million dollars when used on the second decile.[78]

Potential Weaknesses

For USG: A potential social cost of transferable priority review rights is that it could slow down the approval of other equally deserving or more urgently needed drugs in the United States.[79] Additionally, awarding a PRV for successful development and licensure of an MCM does not necessarily ensure that the product will be successfully manufactured. As one SME stated, "A PRV provides an incentive for a company to develop a treatment, but how do we make sure that the product actually gets to people?" Finally, the resource constraints of the FDA must be considered, where reviewers are already working at or near full capacity. As such, the USG and FDA would need to plan and budget

---

[74] Grabowski, 2005.

[75] Effective patent life is defined as the period of patent protection for a drug remaining once the drug is approved by the FDA for marketing.

[76] Grabowski et al., 2007.

[77] Patents granted to pharmaceutical products generally last 20 years, but effective patent life is much shorter because patients are typically granted several years before a product clears the FDA review process.

[78] Matheny et al., 2007; Grabowski, 2005.

[79] Grabowski et al., 2007.

appropriately to be able to absorb the additional workload and accelerated obligations that accompany an expanded PRV program.

For PRV Recipient: The value of the transferable PRV for private pharmaceutical and biotech companies has been debated. Significant uncertainty surrounds the true expected value of the PRV. First, there is no guarantee that the product to which a PRV is applied will attain approval/licensure or that the FDA will respond within 6 months, which limits their value. Additionally, the product to which a company applies the PRV may not gain a sizeable market once approved. Given these uncertainties, our industry experts agree that a PRV alone may not be sufficient to attract large industry participants. However, a combination of PRVs and other incentives could have better success.

Implementation Issues

The implementation of PRVs for biological threats is relatively simple. The path is already in place, such that biodefense products would simply need to be included on the list of eligible candidates. The list of products eligible to receive PRVs, defined as "infectious diseases for which there is no significant market in developed nations and that disproportionally affects poor and marginalized populations," can be expanded by the HHS Secretary as per legislation.[80] Under this legislation, however, this list could not include MCMs to counter chemical and Rad/Nuc agents.

Enhanced Market Exclusivity Rewards

Enhanced market exclusivity rewards ensure that the MCM product is the only one on the market for a period of time that is more extensive than what is currently guaranteed by patents or the 7-year market exclusivity reward under the Orphan Drug Act. The intent of this incentive is to increase returns for MCM development in order to encourage industry participation. It protects intellectual property by providing a legal barrier against would-be competitors from copying the product covered by market exclusivity. During this period, the developer of the MCM effectively has a monopoly and can set price at a level that maximizes profit. While this exclusivity runs concurrently with the

---

[80] SEC. 524. [21 USC § 360n] Priority Review to Encourage Treatments for Tropical Diseases.

regular patent term, it is not at patent, and should remain useful for compounds where natural substances are not eligible for patents on the molecule itself, as well as the dual use of older chemical entities. The provision of a market exclusivity reward is important, as many new and useful therapies are not new molecular entities. For example, the first approved therapy for AIDS in 1987, Zovirax (AZT), was a compound that had previously been investigated as a cancer therapy in the 1960s.[81]

Potential Benefits

For USG: The market exclusivity rewards does not require any upfront payment by the government.[82]

For Market Exclusivity Reward Recipient: The market exclusivity reward lengthens the period in which the MCM retains a monopoly on the market, as well as provides protection to products that are not patentable, such as naturally occurring substances or dual-use products.

Potential Weaknesses

Unless applied to a dual-use product, market exclusivity rewards may not provide adequate incentive for MCM development, since market exclusivity may not be valuable to a product that has no market to begin with. In essence, an MCM may already have exclusivity because it resides in a market that will not likely have competitors in the first place.[83] It is possible that market exclusivity could also de-incentivize innovation and competition in circumstances where one developer is a clear front runner and thus competitors have little hope of capturing any market share.

Implementation Issues

Determining the length of extensions would need to be balanced against the expected R&D costs to the developer and with the expected social value of the new medicine. Additionally, the fact that some MCMs

---

[81] Grabowski, 2005.

[82] Government, through MCM purchases, will likely eventually pay for drugs with prices that are elevated through patent extensions.

[83] Additionally, an MCM would already be very likely to have market exclusivity as an Orphan Drug, so any market exclusivity contract should take that into account.

would already have market exclusivity as orphan drugs may decrease the attractiveness of a market exclusivity reward.

Transferable Patent Extensions

A transferable patent extension allows the patent extension recipient to extend the patent of another product in its pipeline. Specifically, the holder of the patent extension can choose to apply the reward to a valuable blockbuster product, increasing the value of the reward by the profits reaped from the added length of patent exclusivity.

Potential Benefits

For USG: Transferable patent extension has few upfront implementation costs, though government payers such as Medicare or Medicaid will eventually have to pay for the extension through increased reimbursement for the product to which the patent extension is applied.

For Industry: With many large pharmaceutical companies facing patent cliffs on their blockbuster drugs, a patent extension on any of their blockbuster products can add up to several hundred million dollars in revenue.

Potential Weaknesses

A transferable patent extension will likely be immensely valuable for industry. The incentive, however, delays the introduction of generics, thus boosting prices for consumers and payers of the blockbuster drug. It concentrates biodefense costs on a small and seemingly random group of consumers who just happen to be using the product on which the transferable patent extension is placed. Transferable patent extensions are generally viewed as unfair, because they place the burden of MCM development on consumers who would otherwise benefit from the introduction of the generic alternative.[84] Furthermore, private industry owners of the transferable patent extensions would likely place them on their most valuable "blockbuster" patents—the expiry of which would benefit the most consumers.[85]

---

[84] Interviews with SMEs.
[85] Maurer, 2009.

Implementation Issues

Two main considerations exist in implementing transferable patent extensions: (1) the negative political climate surrounding the patent extensions in general, and (2) the calculation of the length of the extension.

Negative Political Climate: Transferable patent extensions are likely to be unpopular amongst legislators for the fairness issues described above. Given the current emphasis on cost containment in the U.S. healthcare system, the proposal also would likely also face stiff opposition from insurance payers and patient groups as well. These types of patent extension schemes were considered, but ultimately rejected by Congress in the debates leading up to the BioShield II legislation in 2006.[86] At this time, prospects of legislative passage for granting patent extensions in exchange for MCM development appear weak.

Calculation of Length of Patent Extension: Determining the length of the patent extension is difficult, and would required careful balancing between the value of the patent extension to the private sector and the social costs associated with administering the incentive. If the patent extension was initiated during the early stages of R&D, then the patent length can be increased to reflect the high levels of uncertainty surrounding the marketability of the product. If the patent extension is granted to products already approved, then the exclusivity period may be shortened extensively to maintain a lower program cost.[87]

Prizes

Prizes are rewards that a sponsor offers to inventors to develop technologies that meet preestablished specifications. Historically, winners were offered cash rewards or project funding, although other rewards such as access to accelerated regulatory reviews (i.e., transferable fast track or priority review vouchers) are certainly possible. Prizes have been used to incentivize innovation in a variety of research-based industries. The Defense Advanced Research Projects Agency (DARPA) Grand Challenge awards cash prizes to makers of driverless vehicles who can design automobiles based on some preset criteria. The National Academy of Engineering's Grainger Challenge offers cash prizes to engineers who can design and build a water treatment system

---

[86] Kremer and Glennerster, 2004.
[87] Grabowski, 2005.

for arsenic-contaminated groundwater in Bangladesh, India. A similar prize mechanism could potentially be used to incentivize the development of MCMs.

Potential Benefits

For Prize Sponsor: Prizes offer several benefits for consideration. Few upfront costs are necessary in administering prizes, and sponsors can expect to only pay when a goal (i.e., a product, technology, animal model, etc.) is successfully developed.[88] As one SME commented, "The great thing about prizes is that we don't have to pick winners and losers up front, and you only have to pay if you get something." Sponsors also do not need to audit or manage research, since developers will be motivated by competition. Once the product is on the market, there need not be a patent monopoly, since the prize sponsor can design the prize to take ownership of the winning patents and subsequently release them into the public domain (i.e., a patent buyout), where competing companies can ensure that the product is competitively priced.[89]

For Prize Recipients: Potential prize recipients generally do not have to deal with extensive government contracting and oversight, which our industry experts have cited as a significant barrier to entry for commercial players into the MCM space. Prizes are also easily legally enforceable, so there is less risk of the government sponsor reneging on any promised payments.[90]

Potential Weaknesses

Our industry experts generally agree that prizes will not induce large pharmaceutical companies to participate in MCM development. They cite that leading pharmaceutical corporations are generally interested in enterprises that have a predictable, sustainable, and profitable market for the MCM product. Part of this skepticism may stem from the fact that prizes usually have a "winner-take-all" approach, where runner-up technologies may be abandoned without ever reaching the market, thereby increasing the product's overall risk profile.[91] Despite these challenges, experiences with DARPA suggest that businesses can

---

[88] Matheny et al., 2007.
[89] Maurer, 2009.
[90] Maurer, 2005.
[91] AdvanceMarkets Working Group, 2005.

respond positively to well-designed prizes.[92] Our subject matter experts also report that innovator biotechs will be more likely to respond to prizes, as opposed to the experienced pharmaceutical companies.

Implementation Issues

Prize designers must consider four main factors—prize size, payment timing, prize specifications, platform technology potential—to ensure an efficient design that will sufficiently incentivize industry players.

Prize size: Setting the size of the prize can be challenging. A prize that is too small may not incentivize MCM development, while a prize that is too large leads to an unnecessary waste of resources that could have been more wisely spent on other endeavors. The prize must be sufficient to compensate for risks of competition, failure, and fault. [93] A prize that induces innovation would need to exceed the private company's opportunity cost of developing the MCM, as well as account for risks of losing the prize to competitors. SMEs suggested asking industry about the size of the reward that would allow them to participate in MCM development efforts. This, however, gives industry an incentive to overstate the size of the prize given that government sponsors oftentimes lack detailed knowledge of actual R&D costs.[94]

Payment timing: The frequency and timing of prize awards must also be considered to ensure that industry participants are properly incentivized. Prizes can be awarded using an end-to-end strategy (E2E), where the prize is a single reward for the entire R&D process, or a pay-as-you-go (PayGo) strategy, where a separate reward is offered for each development substep.[95]

- E2E Prize: Private firms may hesitate to commit hundreds of millions of dollars over fifteen years to pursue an MCM with the uncertain probability of winning a prize in the E2E scenario. To mitigate this risk, a "winner-takes-some" rather than a "winner-takes-all" approach could be used. For example, the E2E prize can be distributed in accordance to each competing product's

[92] DARPA Grand Challenge Website. During the first year, the DARPA Grand Challenge succeeded in attracting 100 teams; by the second year, 195 teams entered the race.
[93] Matheny et al., 2007.
[94] Maurer, 2009.
[95] Terminology from Mauer, 2009.

eventual market penetration.[96] "Winner-take some strategies" can allow multiple products to enter the market, where each drug can offer at least some minor advantage to certain population subgroups.[97] Also, having several products on market will help constrain prices through market competition.[98]

- PayGo: The prize sponsor can create a prize for a small subsection of the R&D process using the PayGo strategy, such as rewarding the first firm to create a successful animal model, or finding a viable candidate for a select target. PayGo prizes work well for many early phase drug development projects, since important information can be elicited and the danger of overpayment is small, as the prize is used to fund only a very limited aspect of the MCM development process.

Prize Specifications. There are also many factors to consider when setting prize rules. If the sponsor has a well-defined idea of the project that they would like to pursue (i.e., identifying a target for an Anthrax therapeutic), they can use "targeted" prizes, or prizes that specify or provide general methods for solving a particular problem. In contrast, a "blue-sky" prize can give contestants the freedom to choose solutions to a particular problem, or even choose the problem. "Blue-sky" prizes work best when the prize sponsor is unclear about the path forward toward a broadly defined goal (i.e., finding improved self-administration technologies for existent vaccines and therapeutics).[99]

Platform Technology Potential. Several of our subject matter experts have commented that big industry players prefer to work on projects where the R&D they are performing has commercial applications, such as testing the addition of adjuvants in vaccines, or designing platform manufacturing capabilities. This approach allows commercial players to access economies of scale for both the development of their commercial products and the MCMs.

Leverageable Capital Asset Investments

The MCME could subsidize or otherwise facilitate the acquisition and development of capital assets (factories, equipment, etc.) for MCM product manufacturing that could subsequently be used for commercial

[96] Maurer, 2005.
[97] Ibid.
[98] Ibid.
[99] Maurer, 2009

products. The purpose of this incentive is to facilitate the acquisition of platform manufacturing technologies to help companies reduce the marginal cost of both the MCM product as well as related commercial application(s).

Potential Benefits

For Industry: Our industry experts agree that access to multiuse manufacturing infrastructure has the potential to induce participation from large pharmaceutical companies. Many players in industry are limited by their current manufacturing infrastructure. Large companies may look toward acquisitions to expand those capabilities, while smaller companies may partner with CMOs. Merck, for example, recently acquired Avecia Biologics,[100] a contract manufacturing firm with specific expertise in microbial-derived biologics, to increase its manufacturing capacity. One SME cited that dual-use MCM manufacturing technologies are particularly attractive because MCM demand will likely not have sufficient volume to warrant full-time use of the facility, allowing any commercial products to be manufactured during downtime.

For USG: Leverageable capital assets are attractive investment opportunities for the USG because they offer a way to provide sustained funding as opposed to a one-time injection of cash. Moreover, reserving or providing large scale manufacturing capacity for companies and academia that lack downstream capabilities helps minimize the time delay in moving MCM discoveries into advanced development. As one SME noted, "It's much better to buy expensive manufacturing than rely on the smaller guys to come up with it on their own."

Potential Weaknesses

Many pharmaceutical companies already have excess capacity in their small molecule facilities, so they may not need additional capital investment for a multiple use facility if the single-use facility is capable of producing the MCM in demand. As one industry representative stated, "If we already have pilot plants, more isn't motivating." The MCME should seek out industry partners that are looking to expand their manufacturing capabilities, or are interested in pursuing R&D of dual-use products with biodefense and commercial indications.

---

[100] Merck Newsroom, 2009.

Implementation Issues

Designing a flexible manufacturing system is difficult. The capital asset owner must demonstrate to the FDA that it can finish a production campaign, and perform a complete cleaning of all instrumentation before starting anew. The process adds another layer of complexity and risk that companies may be averse to taking. However, one industry expert had a much more optimistic view of utilizing facilities for dual-use, stating that equipment used to manufacture a vaccine or therapeutic for an infectious disease should take one to two weeks to clean, and that such a routine is common for large pharmaceutical companies.

MINIMIZING DISINCENTIVES

While implementing incentives can help drive the MCM development engine, it is also important to minimize any disincentives that may obstruct industry participation in MCM development. Efforts to minimize disincentives (or establish "indirect incentives"[101]) can encourage industry participation and thus deserve strong consideration. Below is a brief discussion of approaches to resolving several impediments to industry participation:

- **Regulatory clarity:** The "evolving" regulatory pathway remains unclear to many, particularly with respect to the Animal Rule and its dependence on studies based on new animal models. This added layer of uncertainty creates a considerable disincentive to participation in the MCME. To mitigate this factor, the USG might consider a dedicated Center within the FDA focused on MCMs, with dedicated resources and a unique perspective on public health preparedness and national security. As another option, the USG could stimulate pre-competitive collaboration by creating a PPP to further regulatory science, foster creation of animal models, etc.
- **Sufficient and Sustained Funding:** Perhaps one of the most straightforward paths to minimizing MCME disincentives is to provide predictable and adequate funding of MCM development on both a near term and long term scale. Industry needs to feel confident that the USG is committed to funding a product from R&D through

[101] Scannon, 2008.

to licensure, as well as providing longterm commitments to support sustained procurements in the future. As long as there is uncertainty surrounding the reliability of funding for MCM development, the effectiveness of all other efforts to attract industry engagement will be severely compromised.

- **Contracting flexibility:** The formality of FAR is seen as a burden to many firms considering the prospect of engaging in the MCME. Several interviewees noted that the contracting process could be made more flexible and that the selection of contract types used should be driven by the current stage of development. Approaching the disincentive from a different angle, the USG could consider providing training and support for companies engaging in the contracting process for the first time to accelerate the learning curve.
- **Top-down leadership:** Quite simply, industry needs to see that biodefense is an important mission with solid backing from the White House and Congress. Assembling attractive incentive packages is a bottom-up approach that can only go so far without top-down leadership setting the tone for the criticality of this mission. USG leadership must align and centralize its messaging and efforts across the entire MCME, including HHS, DoD, DHS, and the White House, to neutralize the current perception among interviewees, including industry representatives, that the MCME lacks accountability and doesn't carry the full backing of the Administration. According to one SME, experienced pharmaceutical firms won't engage "unless the president himself asks for it." Such leadership was seen as a necessity by several interviewees to demonstrate the kind of commitment required to make MCM development successful.

CONCLUSIONS AND PATH FORWARD

Selection of Incentives

Interview and document findings suggest the following conclusions regarding potential incentives that could be applied to the MCM business model:

- Pull strategies that increase market sustainability might be more effective than push strategies. One industry interviewee echoed the feeling of many that “the most important incentives are the ones that get us to a market- to product sales.”
- Push strategies should focus on “opening the door to partnerships between companies.”
- No one push or pull incentive will be sufficient to attract companies to the MCME.
- Combinations of push and pull incentives may be sufficient, dependent upon on policy and strategy decisions for each MCM program and on the maturity of the pipelines.
- No one combination of incentives is the “right” model. The effectiveness of different incentives depends on context of policies, products, and players.
- Minimizing disincentives may be enough to “tip the scale” and would “send a signal that the MCME is committed to collaborating with industry.”

Most Frequently Cited Opportunities for Increasing Participation

Under the current policy of focusing on post-event response based on stockpiled product, opportunities to increase industry participation exist across the latter three segments of the MCME Business Model Framework. Based on the literature review and the perspectives of SMEs collected throughout interviews, we provide below a summary of the most frequently cited opportunities for increasing the level and mix of involvement by pharmaceutical and biotechnology companies the MCME.

These opportunities assume a continued policy of focusing on post-event response based on stockpiled product. Alternative policies could have a cascade of alternate approaches.

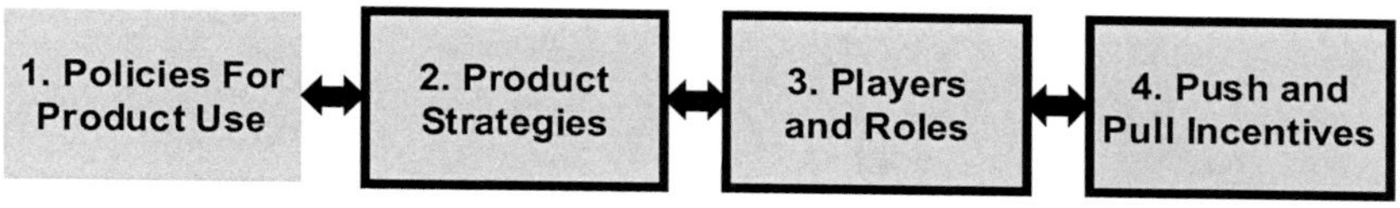

FIGURE E-4 Opportunities for inducing industry participation stem from product strategies, players and roles, and push and pull incentives.

Product Strategies: To date the pursuit, broad-spectrum approaches within the MCME product strategy have been limited. Successful efforts have primarily been constrained to new indications for antibiotics. In order to expand into the range of competitive returns for industry, it will be necessary to increase the emphasis placed on dual-use products, platforms, and technologies with commercial applications. Examples of this product strategy include vaccine platforms and multiple indications on a single drug product.

Players and Roles: To date, the MCME's business model has primarily attracted innovator companies to manage the entire development chain, from early stage discovery research through late stage development and production. A more effective approach may be to focus on taking advantage of the strengths of a broader set of industry players. Specifically, the USG should look to innovator companies to drive products through proof of concept, while relying on larger, more experienced companies for late-stage development and manufacturing. This division of labor caters to the strengths and risk profiles of each group. To that end, the USG and MCME leaders should explore opportunities to promote partnering throughout the enterprise landscape. The concept of partnerships, in multiple forms, was consistently recommended by stakeholders and subject matter experts we interviewed.

Incentives: A variety of potential incentives were explored throughout the literature and stakeholder interviews. The single most critical factor in the eyes of the experts we surveyed was that the USG must demonstrate a reliable, long-term market for MCM products. Whether accomplished by increasing/extending BioShield, or through some other approach, establishing market credibility is pivotal in positioning MCME projects for serious consideration by industry.

Additional incentives that drew significant attention and favor among the interview candidates and literature authors include the following:

- Priority Review Vouchers—PRV's appeal to a broad range of stakeholders because they represent an incentive that could be highly valued by industry while carrying limited social cost.
- Tax incentives—This tool could be particularly effective if structured such that larger companies that partner with innovator companies earn a valuable tax credit. Such an incentive would promote the partnerships noted in the previous section as a powerful approach to development.
- Additional Funding—Biopharmaceutical research is expensive. With $5.6 billion allocated over 10 years to purchase at least 14 CBRN threat agents, BioShield is unlikely to draw participation from large, pharmaceutical companies.[102] No matter how funds are shared among push and pull mechanisms targeted at pharmaceutical companies for MCM research, it is unlikely that $900 million dispersed among various incentives would appear more attractive to industry than a $3 billion commercial drug.[103]

APPENDIXES

Examples of Incentive Structures from Other Sectors

Renewable Energy

- Goal: Encourage and increase adoption of renewable energy sources despite higher costs/risks to suppliers
- Incentives: Guaranteed capacity, risk reduction, and adjusted compensation
- Relevance: Similar to MCME's high opportunity cost and expensive manufacturing processes

---

[102] Matheny et al., 2007.
[103] Ibid.

Commercial Space Launch

- Goal: Attract first rate talent and technology to an extremely high risk/high cost industry
- Incentives: Prizes, pay-for-performance, leverageable capital investments
- Relevance: Similar to MCME's "lumpy" revenue streams, long lead times, exclusive government market

Defense

- Goal: Encourage industry to support a market that is almost exclusively government driven; encouraging production and capacity investment during difficult economies
- Incentives: Demand aggregation, purchase commitments, subsidies, knowledge sharing
- Relevance: Mirrors MCME's need for multiple incentive approaches

Semiconductors

- Goal: Generate rapid innovation and advancement in U.S. semiconductor industry to reclaim lost market share
- Incentives: Trade agreements, public–private partnership (SEMATECH)
- Relevance: Provides excellent model of public–private partnership success

Nuclear Technologies

- Goal: Stimulate interest in the nuclear power industry where liability, in the case of a nuclear accident, was a serious obstacle
- Incentives: Liability protection through the Price Anderson Act, which provided more than $9.5B in insurance coverage
- Relevance: Like MCME, concern over indemnification hinders industry participation

Biotechnology Incubators

- Goal: Foster growth of innovative technologies and ideas from start-up ventures who cannot reach commercialization alone

- Relevance: Provide services in line with needs of inexperienced biodefense companies
- Incentives: Provide services to expedite R&D of promising biotechnologies

BIBLIOGRAPHY

AdvancedMarkets Working Group. 2005 (April 7). *Making markets for vaccines.* Center for Global Development.

Andre, F. E. 2002. How the research-based industry approaches vaccine development and establishes priorities. *Developments in Biologicals* 110:25-29.

Audretsch, D. B., B. Bozeman, K. L. Combs et al. 2002. The economics of science and technology. *Journal of Technology Transfer* 27:155-203.

Baker, D. 2004 (Sept. 22). *Financing drug research: What are the issues?* Center for Economic and Policy Research.

BARDA Authorities Assessment. 2008 (Sept. 12). HHS Transition Package.

Berndt, E., and J. Hurvitz. 2005. Vaccine advance-purchase agreements for low-income countries: Practical issues. *Health Affairs* 24(3):653-665.

Berndt, E., et al. (incl. R. Levine). 2006 (Aug.). *Advance market commitments for vaccines against neglected diseases: Estimating costs and effectiveness.* Center for Global Development Working Paper No. 98.

Brogan, D., and E. Mossialos. 2006 (Aug.). Applying the concepts of financial options to stimulate vaccine development. *Nature Reviews Drug Discovery* 5:641-647.

Challenge and Opportunity on the Critical Path to New Medical Products. 2004 (March). FDA Critical Path Initiative.

Chesbrough, H. 2003. *Open innovation: The new imperative for creating and profiting from technology.* Harvard Business School Publishing Company.

Commercial Spaceflight Federation. 2009 (June 29). *Commercial spaceflight in low earth orbit is the key to affordable and sustainable exploration beyond.* Input into the Review of U.S. Human Space Flight Plans Committee.

Critical Path Opportunities Report. 2006 (March). FDA Critical Path Initiative.

CSC News Release. 2008 (June 5). *CSC's DynPort Vaccine Company to continue plague vaccine development.*

DARPA Grand Challenge Website. (accessed March 7, 2010).

DiMasi, J. A., and H. G. Grabowski. 2007. The cost of biopharmaceutical R&D: Is biotech different? *Managerial and Decision Economics* 28:469-479.

DiMasi, J., et al. 2010 (Feb. 3). Trends in risks associated with new drug development: Success rates for investigational drugs. *Nature Publishing Group: Clinical Pharmacology and Therapeutics.*

Doremus, P. 2008 (July 17). *The R&E tax credit: Rationale, structure and performance.* The National Institute of Standards and Technology, Strategic Planning & Economic Analysis Group.

Fuerst, T. et al. 2009. *Ensuring biologics advanced development and manufacturing capability for the United States Government—A summary of key findings and conclusions.* Pittsburgh, PA: UPMC.

GAVI Presentation. *Accelerating the development and use of new vaccines: Economic and financial incentives.*

GAVI Press Release. 2009 (Oct. 12). *Update: Pneumococcal AMC receives four offers from vaccine manufacturers.* http://www.vaccineamc.org/updateoct12_09.html.

Global Health Security Initiative Website. 2010. (accessed March 7, 2010).

Global Security Newswire. 2010 (Feb. 26). *Former HHS official backs home kits for biodefense.*

Gottron, F. 2007 (June 11). *Project BioShield: Appropriations, acquisitions, and policy implementation issues for Congress.* Washington, DC: Congressional Research Service.

Gottron, F., and E. Fisher. 2004 (Aug. 27). *Project BioShield: Legislative history and side-by-side comparison of H.R. 2122, S. 15, and S. 1504.* Washington, DC: Congressional Research Service.

Grabowski, H. 2005 (May/June). Encouraging the development of new vaccines. *Health Affairs* 24(3):697-700.

Grabowski, H. 2005. Increasing R&D incentives for neglected diseases: Lessons from the Orphan Drug Act. In Cambridge University Press conference volume, edited by J. Reichman. Pp. 457-480.

Grabowski, H., D. Ridley, and J. Moe. 2009. Encouraging innovative treatment of neglected diseases through Priority Review Vouchers. *Prescribing Cultures and Pharmaceutical Policy in the Asia-Pacific.* Pp. 347-365.

Gronvall, G., et al. on behalf of the Alliance for Biosecurity. 2007. The FDA Animal Efficacy Rule and biodefense. *Nature Biotechnology.*

Guidance for Industry: Animal models—Essential elements to address efficacy under the Animal Rule. 2009 (Jan.). HHS, FA, CDER, CBER.

Hatchett, R. 2009 (Oct. 27). *Alternative models for medical countermeasures development.*

HHS PHEMCE stakeholders workshop 2007 report. 2008 (July/Aug.). Biomedical Advanced Research and Development Authority.

Higgins, M., and S. Graham. 2009 (Oct.). Balancing innovation and access: Patent challenges tip the scales. *Science* 326(5951):370-371.

Hsu, J., and E. Schwartz. 2003 (Oct.). *A model of R&D valuation and the design of research incentives.* NBER Working Paper 10041.

Hu et al. 2007 (March 12). *The innovation gap in pharmaceutical drug discovery and new models for R&D success*. Kellogg School of Management.

Inglesby, T. 2008. *Building an effective government–private sector partnership for biodefense.* PHEMCE Stakeholders Workshop.

IOM (Institute of Medicine). 2002. *Protecting our forces: Improving vaccine acquisition and availability in the U.S. military* Washington, DC: The National Academies Press.

IOM 2004. *Giving full measure to countermeasures: Addressing problems in the DoD program to develop medical countermeasures against biological warfare agents*. Washington, DC: The National Academies Press.

IOM. 2008. *Addressing the barriers to pediatric drug development:. Workshop summary*. Washington, DC: The National Academies Press

IOM. 2009. *Breakthrough business Models: Drug development for rare and neglected diseases and individualized therapies*. Washington, DC: The National Academies Press.

IOM. 2010 (Feb.) *The Public Health Emergency Medical Countermeasures Enterprise: Innovative strategies to enhance products from discovery through approval*. An Institute of Medicine Workshop.

Klotz, L. C., and A. Pearson A. 2009. BARDA's budget. *Nat Biotechnol* 27(8):698-699; author reply 699.

Kremer, M. 2000 (May 24). *Creating markets for new vaccines, Part 1: Rationale*. The Brookings Institute and NBER.

Kremer, M. 2000 (May 24). *Creating Markets for New Vaccines*. Harvard University, Brookings Institute, and NBER.

Lightfoot, N. 2009 (Nov.). *The Global Health Security Initiative (GHSI) Public Health Emergency Medical Countermeasures (MCM) Workshop.* U.S. Department of Health and Human Services.

LINUX Website. (accessed March 7, 2010).

Matheny, J., et al. 2006. Drug and vaccine development for infectious diseases: The value of Priority Review Vouchers. *Clinical Pharmacology and Therapeutics* 85(6):571-572.

Matheny, J., et al. 2007. Incentives for biodefense countermeasure development. *Biosecur Bioterror* 5(3):228-238.

Matheny, J., M. Mair, and B. Smith. 2008. Cost/success projections for US biodefense countermeasure development. *Nat Biotechnol* 26(9):981-983.

Maurer, S. 2005. *The right tools: Designing cost-effective strategies for neglected disease research*. Goldman School of Public Policy, University of California–Berkeley.

Maurer, S. 2009. Squeezing value from Homeland Security research: Designing better R&D incentives. In *WMD Terrorism: Science and Policy Choices*. The MIT Press.

McKenna, F. 2009 (July 20). *Commercial launch faces economic reality, again.* From International Launch Services. *Space News*.

Merck Newsroom. 2009 (Dec. 27). *Merck & Co. to acquire Avecia Biologics*.

Milken Institute. 2006 (Oct.). *Financial innovations for accelerating medical solutions*. Milken Institute Publication.

Monath, T. *Developing products against biological and emerging threats: Progress, pitfalls and set-backs*. KPCB Presentation.

Moran, M. 2005. A breakthrough in R&D for neglected diseases: New ways to get the drugs we need. *PLoS Medicine* 2(9):0828-0832.

Moran, M. 2006 (Sept.). *The new landscape of neglected disease drug development.* Medicines for Malaria Venture.

Munos, B. 2006 (Aug. 18). Can open-source R&D reinvigorate drug research? *Nature Reviews Drug Discovery; Perspectives.*

Munos, B. 2009. Lessons from 60 years of pharmaceutical innovation. *Nature Reviews Drug Discovery*. pp. 959-968.

Munos, B., and W. Chin. 2009 (Dec.). A call for sharing: Adapting pharmaceutical research to new realities. *Science Translational Medicine* 1(9).

Optimizing industrial involvement with medical countermeasure development: A report of the National Biodefense Science Board. 2010 (Feb.). Department of Health and Human Services.

Orphan Drug Act. 1983. Public Law 97-414.

Parker, G. 2007. HHS Public Health Emergency Medical Countermeasures Enterprise implementation plan for chemical, biological, radiological and nuclear Threats. *Federal Register* 72(77):20117-20128.

PHEMCE Strategy and Implementation Plan. 2007. Department of Health and Human Services, Assistant Secretary for Emergency and Preparedness Response.

PHEMCE Workshop 2008 & BARDA Industry Day Report. 2008 (September). Biomedical Advanced Research and Development Authority.

Plahte, J. 2005 (Jan.). *Tiered pricing of vaccines: A win-win-win situation, not a subsidy.* Lancet Infect Dis 5:58-63.

Radcliffe, S. 2008. *Strategies and suggestions: Developing and sustaining a biodefense industry.* PHEMCE Stakeholders Workshop.

Report on Biological Warfare Defense Vaccine Research & Development Programs (Top Report). 2001. Department of Defense.

S. 3678: Pandemic and All-Hazards Preparedness Act. 2005-06. 109th Congress.

Scannon, P. 2008 (June). *A realistic path to future medical countermeasure development and acquisition for the Strategic National Stockpile.*

Scannon, P. 2008. *Strategies and suggestions for developing and sustaining a biodefense industry: A focus on medical countermeasures*. PHEMCE Stakeholders Workshop.

Schacht, W. 2005 (May 6). *Bioterrorism countermeasure development: Issues in patents and homeland security*. CRS Report to Congress.

Schacht, W., and J. Thomas. 2007. *Bioterrorism countermeasure development: Issues in patents and homeland security*. Washington, DC: Congressional Research Service.

Schoonmaker, M. 2005 (March 9). *From bench to bedside: The role of Health and Human Services (HHS) in the evaluation of new medical products*. CRS Report to Congress.

Touse, A., and H. Kettler. 2004 (March 9). Advance price or purchase commitments to create markets for treatment for diseases of poverty: Lessons from three policies. *Bulletin of the World Health Organization* 84(4).

Tucker, J. 2009. Developing medical countermeasures: From BioShield to BARDA. *Drug Development Research* 70(4):224-233.

White House's Homeland Security Presidential Directive 18. 2007 (Feb. 7). White House Office of the Press Secretary.

Witty, A. 2010 (Jan.). *Open labs, open minds: Breaking down barriers to innovation and access to medicines and vaccines in the developing world*. From GlaxoSmithKline—Remarks delivered to the Council on Foreign Relations, New York.

Zerhouni, E. 2009 (Dec.). Peer-to-peer sharing spurs scientific innovation. *Science Translational Medicine* 1.

LIST OF INTERVIEWS

The following is a list of individuals who were interviewed by PRTM for the purpose of this White Paper. Affiliations listed reflect the individuals primary association as of the date of the interview.

Jennifer B. Alton, *Bavarian-Nordic*
Marguerite Baxter, *Novartis*
Ernst Berndt, *MIT Sloan School of Management*
Luciana Borio, *HHS/FDA*
Carla Botting, *Malaria Vaccine Initiative*
Mike Callahan, *DARPA*
Francesca Cook, *PharmAthene*
Jodie Curtis, *Alliance Secretariat*
Jeffrey Fu, *Merck*
Thomas Fuerst, *HHS/ASPR/BARDA*
John Grabenstein, *Merck; National Biodefense Science Board*
Henry G. Grabowski, *Duke University*
Maureen Hardwick, *Alliance for Biosecurity*
Richard Hatchett, *HHS/NIH/The White House*
Robert House, *DynPort Vaccine Company*
Gerry Kovacs, *HHS/ASPR/BARDA*
Michael Kurilla, *HHS/NIH*
Carol Linden, *HHS/ASPR/BARDA*
Boris D. Lushniak, *HHS/FDA*
Ronald Manning, *HHS/ASPR/BARDA*
Jerry Middleton, *Elusys Therapeutics*
Thomas P. Monath, *Kleiner Perkins Caufield & Byers*
Bernard Munos, *Eli Lilly*
David Noll, *Bavarian-Nordic*
Linda Nyari, *PATH*
Joanna Prasher, *HHS/ASPR/BARDA*
Rino Rappuoli, *Novartis*

John H. Rex, *AstraZeneca*

David Ridley, *Duke University*

Robin Robinson, *HHS/ASPR/BARDA*

Eric Rose, *Siga Technologies*

Phillip Russell, *Sabin Vaccine Institute*

Jerry Sadoff, *AERAS*

Patrick J. Scannon, *Xoma; National Biodefense Science Board*

Bradley Smith, *UPMC Center for Biosecurity*

CONTACT INFORMATION

Michael Mair, MPH
Program Analyst
Policy, Planning, and Requirements Division
Office of the Assistant Secretary for Preparedness and Response
Department of Health and Human Services Michael.Mair@hhs.gov
202-260-1343

James Guyton
Principal
PRTM Management Consultants, LLC
1750 Pennsylvania Avenue NW
Suite 1000
Washington, DC 20006
jguyton@prtm.com
410-428-3259

Hannah McClellan
Associate
PRTM Management Consultants, LLC
1750 Pennsylvania Avenue NW
Suite 1000
Washington, DC 20006
hmcclellan@prtm.com
617-680-4500

Fanni Li
Consultant
PRTM Management Consultants, LLC
1750 Pennsylvania Avenue NW
Suite 1000
Washington, DC 20006
fli@prtm.com
781-526-8191